D0979503

THE TRUTH ABOUT CANADA

BY THE SAME AUTHOR

The Betrayal of Canada

A New and Better Canada

At Twilight in the Country

Pay the Rent or Feed the Kids

The Vanishing Country

Rushing to Armageddon

THE
TRUTH
ABOUT CANADA

Some Important, Some Astonishing,

and Some Truly Appalling Things

All Canadians Should Know

About Our Country

MEL HURTIG

EMBLEM

McClelland & Stewart

Copyright © 2008 by Mel Hurtig

Cloth edition published 2008
Emblem edition published 2009

Emblem is an imprint of McClelland & Stewart Ltd.
Emblem and colophon are registered trademarks of McClelland & Stewart Ltd.

All rights reserved. The use of any part of this publication reproduced, transmitted
in any form or by any means, electronic, mechanical, photocopying, recording, or
otherwise, or stored in a retrieval system, without the prior written consent of the
publisher – or, in case of photocopying or other reprographic copying, a licence from
the Canadian Copyright Licensing Agency – is an infringement of the copyright law.

Library and Archives Canada Cataloguing in Publication

Hurtig, Mel
        The truth about Canada : some important, some astonishing, some truly
appalling things all Canadians should know about our country / Mel Hurtig.

Includes bibliographical references.

ISBN 978-0-7710-4166-2

        1. Canada – Politics and government – 2006-. 2. Canada – Economic
conditions – 1991-. 3. Canada – Social conditions – 1991-. 4. Health status
indicators – Canada. 5. Canada – History – 21st century. I. Title.

FC640.H87 2009   971.07'3   C2008-904242-5

We acknowledge the financial support of the Government of Canada through the
Book Publishing Industry Development Program and that of the Government of
Ontario through the Ontario Media Development Corporation's Ontario Book
Initiative. We further acknowledge the support of the Canada Council for the Arts
and the Ontario Arts Council for our publishing program

Typeset in Fairfield by M&S, Toronto
Printed and bound in Canada

A Douglas Gibson Book

McClelland & Stewart Ltd.
75 Sherbourne Street
Toronto, Ontario
M5A 2P9
www.mcclelland.com

2   3   4   5   13   12   11   10   09

The Truth About Canada is dedicated
to the wonderful people at Unicef,
the United Nations Children's Fund.

Ultimately, this book is about social justice, the quality of life, and about democracy. The following quote is from Unicef's 2007 *Innocenti Report Card* 7, originating at the Innocenti Research Centre in Florence, Italy. It sums up some of the key chapters that follow.

The true measure of a nation's standing is how well it attends to its children – their health and safety, their material security, their education and socialization, and their sense of being loved, valued, and included in the families and societies into which they are born.

TABLE OF

CONTENTS

ACKNOWLEDGEMENTS

With special thanks to my publisher, Douglas Gibson, and to Philip Cross, Jillian Skeet, Geoffrey Stevens, Julie Maloney, Willem Hubben, David Green, Martha Friendly, Jared Bernstein, Carla Curran, Pierre Lalonde, Larry Gordon, Andrew Sharpe, Kim Kierans, Rob Macintosh, Carol Goar, Dennis Raphael, Mary Griffiths, Peter C. Newman, David Hughes, Mark Jaccard, Geoff Ballinger, Jim Stanford, Michael Veall, Duncan Cameron, Richard Chaykowski, Paul Axelrod, Richard Stursberg, Ian Morrison, Lawrence Martin, Peter Desbarats, Stewart Taylor, David Sabourin, Les Shinder, Thomas Walkom, Michael Byers, Eric Reguly, Steve Staples, John Ibbitson, Gordon Laxer, Barry Mersereau, Wendy Holm, Bill Potter, Neil Brooks, Bill Tielman, Janine Brodie, Aaron Paton, David Morley, Peter Julian, Cathy Oikawa, Nigel Fisher, Jeffrey Simpson, James Travers, Erin Weir, Doug Pepper, Roy MacSkimming, Roger Jullion, Allan Tomas, Kelly Patterson, Andrew Jackson, Owen Adams, Alain Rineau, Don Drummond, Rob Rainer, Paul Cappon, Michael McBane, John Ralston Saul, Mel Clark, Walter Dorn, David Crane, Nicholas Heap, Michael Goldberg, Jeff Merrett, and John Godard.

Also, special thanks are due to the wonderful people at Statistics Canada and at the OECD who have been so very helpful in steering me to the numbers and the sources I needed in writing this book.

PREFACE

This book is about how Canada has changed, and changed very much for the worse, under the governments of Brian Mulroney, Jean Chrétien, Paul Martin, and Stephen Harper. It is also about how, as a result of the profound changes that have occurred, we are no longer the country we think we are, and no longer the people we think we are.

In chapter after chapter, you will discover just how very different we've become from our long-time self-image and from what has been our international image. You'll see how far we've departed from the principles and the ideas that helped Canada become one of the most admired countries in the world and the country the overwhelming majority of Canadians have so cherished for so long.

An important feature of *The Truth About Canada* is the fascinating international comparisons it contains that show how we stack up against other countries around the world, but principally against the other OECD developed countries. It's no exaggeration to say that you will find a great many of these comparisons disappointing, shocking, and even appalling.

Another main theme in the pages that follow is the dismal failure of our powerful corporate leaders to use their gigantic, record-breaking profits and reduced taxes to adequately invest in our country and to conduct reasonable levels of research and development that would help make Canada more innovative, more productive, and more competitive so we can raise our overall standard of living. You will find many of the facts that follow relating to big business in Canada both disturbing and dismaying.

A further theme is the unparalleled sellout of our country in a manner no other developed country would ever dream of allowing. While this has been taking place at an accelerating rate, the purposeful dissemination in the print media of false information about rapidly growing foreign ownership and control of Canada goes a long way towards explaining why our myopic politicians have failed to take action on this and other related problems that are very quickly robbing us of our ability to plan and manage our own future.

The chapter on the Free Trade Agreement is subtitled "The Most Colossal Con Job in Canadian History." When you read it, I hope you will ask yourself why you have never read any of this information in our newspapers or magazines, or have never seen anything remotely similar on television. God knows, you've been inundated with an abundance of right-wing, continentalist propaganda to the contrary. The chapter on the media in Canada should help explain why this important information has never before been available to you.

When you read the economic chapters in this book, on foreign ownership, trade, investment, productivity, competitiveness, and taxation, I hope you will be aware of the fact that exactly the same people who have left us in these weakened positions have for some time been very secretly planning more of the same in private, high-level meetings designed to integrate Canada further into the United States.

Big business, in the form of the Business Council on National Issues and its well-financed successor, the Canadian Council of Chief Executives (the very same people who helped put Canada, as you will see, into two terrible so-called "trade" agreements), are now covertly planning "deep integration" with the United States, a process that will rob us of our ability to maintain our independence, protect our sovereignty, and preserve the important values so many of us cherish.

I hope you will be angry after reading *The Truth About Canada*, very angry. Angry at greedy, hypocritical, intentionally misleading corporate executives, and angry at the remarkably inept politicians who have allowed a small and wealthy plutocracy to sell out our country and our destiny for their own selfish motives.

The Truth About Canada is the result of many long days, months, and years of research. It certainly will be regarded as my most controversial book, and will bring immediate cries of protest from the usual Neanderthals at the Fraser Institute, the C.D. Howe Institute, the CCCE, the increasingly continentalist Conference Board of Canada, and, of course, the house organ of all of them, the *National Post*.

One editor asked me if I was not apprehensive about the strong criticism such a tough book will inevitably bring. The answer is simple. You cannot ever expect to accomplish anything important without bringing criticism from the entrenched forces this book describes, criticizes, and blames for what has gone wrong in our country.

I have been very fortunate in having some of the best minds in the country available to me for consultation as I wrote *The Truth About Canada*. You will find their names on the acknowledgements page. Many of the most important pages of original research in this book are the result of their help, for which I am very grateful.

Whether it's our pathetically low number of doctors, our high comparative levels of both adult and child poverty, our truly awful record of educational funding, our shameful levels of foreign aid and peacekeeping, our abysmal voter turnout comparisons, our totally inadequate research and patent performances, our high infant and under-five mortality rates, the broad deterioration in our social programs, our increasing gaps in distribution of income and wealth in Canada, our treatment of our aboriginal peoples, the rapid decline of our manufacturing sectors, our serious post-secondary education problems, our continuing and very dangerous decentralization, our coming confrontation with the United States over water, our mind-bogglingly stupid NAFTA agreements regarding oil, natural gas, and water – in any or all of these topics, and in many more, you will frequently encounter vitally important and newly documented information that will make you cringe.

In researching this book, I found it truly dismaying to see how often the print media in Canada totally ignores or distorts information that doesn't fit with their own philosophical/editorial positions. Quite often I pick up a morning paper and read about a new Statistics Canada, OECD, or other release, knowing that the newspaper story and the documented reality are vastly different from one another. The chapters in this book on poverty, social policy, wages, distribution of income and wealth, profit and investment, taxes, foreign ownership, the FTA and NAFTA, government in Canada, decentralization, and energy policy all contain information you likely never have seen before or information that runs contrary to the conventional and widely accepted opinions in Canada found in much of our media and intentionally promoted by our far-right "think-tanks," business leaders, and politicians.

I hope that when you read the following pages you will tell others about what you have read. If you need more information about sources, or if you want to be added to my e-mail list (I sometimes send out four or five items a month), you can contact me at melhurtig@shaw.ca. If required, I will get back to you as soon as I can, but please understand that this will likely take quite some time.

Mel Hurtig
Vancouver
December 2007

PART ONE

HEALTH CARE IN CANADA AND OUR TRAGIC, INEXCUSABLE SHORTAGE OF DOCTORS

AN ABSURD SITUATION NO ONE WANTS
TO TAKE RESPONSIBILITY FOR

Surely one of the most startling facts you will find in this book is this: during the years 1990 to 2004, in terms of the number of physicians per 100,000 people, Canada stood far, far down the list of all countries, in an appalling 54th place, according to the United Nations's Human Development Report 2007/2008. With only 214 doctors per 100,000 during those years, we were down among some of the poorest and least developed countries in the world, and far below most other countries in the OECD (Organization for Economic Co-operation and Development).

Some comparisons in physicians per 100,000: Cuba 591, the United States 549, Belgium 449, Estonia 448, Greece 438, Russia 425, Italy 420, Turkmenistan 418, and Georgia 409. All the following countries have over 300 physicians per 100,000: Norway, Iceland, Sweden, Switzerland, the Netherlands, Finland, Austria, France, Spain, Germany, Israel, Portugal, the Czech Republic, Malta, Hungary, Argentina, Lithuania, Slovakia, Uruguay, Latvia, Bulgaria, Lebanon, Kazakhstan, Armenia, and Azerbaijan.[1]

How could this be? How could one of the world's wealthiest developed countries have found itself in such an absurd position? And how is it that, in the millions of words in the press and from our politicians about the problems of the Canadian health-care system, this preposterous shortage of physicians has received so little attention?

One very prominent and widely respected physician, who has to go unnamed in this book, told me that "the poor planning in health human resources is one of the most serious public policy failings that our governments have ever been involved in, yet no one wants to take responsibility."[2]

In the 1970s, Canada had the second highest physician-to-population ratio among developed countries. But subsequent ill-considered actions by provincial governments intentionally reduced the number of doctors in medical schools across the country.

Today, despite our acknowledged alarming shortage of doctors – and despite the large number of young men and women anxious for admission to our medical schools, even with high marks and excellent resumés – large numbers of our very best and brightest young people who want to be doctors are being forced out of the country. At the same time, some Canadian universities have raised annual medical-school tuition fees to astronomical levels. At the University of British Columbia they went up 375 percent in four years, to $14,280 in 2006, and only 16 percent of applicants were accepted.[3] At the University of Toronto, fees were just under $17,000.

While a recent federal budget committed to increasing medical-school enrolment by 15 percent , that will still leave Canada far short of the number of doctors needed, and the increased number of new graduates will not be available until 2011 at the earliest. Meanwhile, in a 2007 survey, over two million Canadians said they had tried but failed to find a family doctor during the previous year, and wait times in this country are far longer than in Germany, New Zealand, Australia, Britain, and a long list of other countries.

How did this happen? In the early part of the 1990s, a consensus developed that there was a surplus of doctors in Canada and that steps had to be taken to control the growth in the number of physicians. In 1991, the Barer-Stoddart Report suggested a 10 percent reduction in the number of Canadian medical students. The next year, a conference of provincial and territorial ministers of health agreed to reduce undergraduate medical enrolment by 10 percent, starting in 1993. They also agreed to reduce the reliance on international medical graduates, reduce postgraduate trainees, and maintain or reduce physician-to-population ratios.

In 1985, medical-school enrolment in Canada was 1,835. By 1999, it had dropped to 1,516. In 1995, there were 191 physicians per 100,000 population. By 2021, this is projected to drop to 130 per 100,000, ridiculously inadequate for patient requirements. By 2005, the average graduation rate for doctors in the OECD was 34 per 1,000 practising doctors, "too low to meet the expected increase in demand." In Canada, it was closer to only 25 per 1,000 practising doctors.[4]

Between 1993 and 2003, Canada's population increased by 13 percent, but the number of doctors declined by 5 percent. On average, Canada now has about a third fewer doctors per population than other OECD countries and only slightly more than half as many per population as in France, Germany, or the United States. Despite the belated recognition of our serious problem, Canada is still well down the list, in 26th place, in the annual growth rate in the number of practising physicians during the period 1990 to 2003.

In 2005, there was a provincial high of 218 physicians per 100,000 population in Nova Scotia, 218 in Quebec, 205 in Yukon, and 199 in British Columbia. At the other end of the scale, there were only 46 in Nunavut, 103 in the Northwest Territories, 144 in Prince Edward Island, and 156 in Saskatchewan. In the middle, there were 193 per 100,000 in Newfoundland and Labrador, 188 in Alberta, 179 in Manitoba, 176 in Ontario, and 172 in New Brunswick. (Note how very low these numbers are compared to the numbers for the most developed countries that are listed in this chapter's second paragraph.)

And supposing you don't live in a big city? In 2004, over 21 percent of Canadians lived in rural areas. The same year, only about 9.4 percent of all physicians practised in rural areas.

One important positive: In 2004, for the first time in 40 years, more Canadian doctors returned to Canada than the number who left. In 2005, only 186 doctors left for the United States, but the number returning to Canada was up to between 200 and 300.[5] While it remains true that about one in nine trained-in-Canada doctors is now practising medicine in the United States (over 8,100 in 2006), the tide has definitely been turning, and there are signs that many more Canadian doctors in the

United States are unhappy and are considering returning to Canada. Is it any wonder that there is growing disenchantment among Canadian doctors in the United States, given the increasing dominance of the widely disliked Health Maintenance Organizations (HMOs) and the growing cost of malpractice insurance (up to as much as $30,000 a month)? More than a few doctors in the United States have been turning their backs on their own profession.

I find it difficult to understand why we give medical students such a quality, heavily subsidized, lengthy education without requiring them to practise for at least 10 years in Canada, or, if they wish to emigrate sooner, to return the large subsidies they have received to the universities or hospitals where they got their education.

Is the amount we pay our doctors part of the problem? In a list of 19 OECD countries, Canadian specialists are not as well paid as specialists in Belgium, the Netherlands, or the United States. But they receive better remuneration (as a ratio of GDP per capita) than specialists in all the other 15 countries. The comparative picture is somewhat similar for general practitioners, where Canadian GPs are at 3.45 percent of GDP, behind only the United States at 4.18 percent, the Netherlands at 3.73 percent, and Germany at 3.61 percent, and once again they are better paid than the GPs in the 15 other countries.[6]

For those who think that the amount we pay doctors is the major reason for escalating health-care costs, it's interesting to note that from 1984 to 1992 physicians not directly employed by hospitals or public-sector health agencies accounted for between 15 and 15.7 percent of health expenditures in Canada, but by 2006 this had fallen to 13.1 percent.

Now, to see what we may well expect in the future, let's turn our attention to the Canadian Medical Association (CMA). In two votes at their annual convention in Charlottetown in August 2006, Canadian doctors supported a two-tier health-care system and rebuffed the efforts of their colleagues who wanted the CMA to curb private medical insurance and/or doctors moonlighting by offering their services both publicly and privately, the latter to patients willing and able to pay out of their own pockets. In the words of Dr. Danielle Martin, head of Canadian

Doctors for Medicare: "The CMA has now marginalized itself. It has shown that it's out of touch with the majority of its members and certainly the majority of Canadians. . . . This is a real blow to the credibility of the profession."

Then, in July 2007, the CMA, to the dismay of many in the profession, came out strongly in favour of a two-tier health-care system that would allow doctors to work simultaneously in both the public and private systems, a proposal that would inevitably lead to queue-jumping and a deterioration of public health care. In most countries, doctors are prohibited from practising in both the public and private sectors. Stephen Harper had it right when he wrote to Alberta premier Ralph Klein, "Dual practice creates conflict of interest for physicians as there would be a financial incentive for them to stream patients into the private portion of their practice." And federal health minister Tony Clement said, "How can you improve access if doctors are spending part of their time inside the system and part of their time outside the system? No one has shown how that can be done."[7]

Gordon Guyatt, professor of medicine at McMaster University, was direct and to the point in his response to the CMA's proposal: "The CMA is acting on the basis of self-interest instead of the public interest. For-profit clinics would not lead to the training of a single additional doctor or nurse. Indeed, such clinics would suck desperately needed personnel from not-for-profit hospitals and clinics."[8]

Wendy Fucile, in the *Toronto Star*, wrote:

> Nurses reject CMA's recipe and say it is a privatization gimmick. In an open letter to Prime Minister Stephen Harper, the Registered Nurses' Association of Ontario urged the PM to restate, in no uncertain terms, that physicians will not be allowed to practise simultaneously in both the private and public health-care systems.
>
> A wealth of evidence shows that allowing physicians to practise in both public and private systems decreases access

to health care, costs taxpayers more, and results in lower quality of care – including higher rates of complications and deaths. A parallel private system siphons health-care professionals and drains resources out of the public system.

Countries with parallel private hospitals have larger and longer waiting lists in their public hospitals. A parallel private system allows for-profit clinics and the physicians who work in them to benefit from people's vulnerability in times of illness. These clinics cherry-pick patients who are healthier, younger or have conditions that are cheaper to treat, leaving more complicated cases to a public system with fewer health-care professionals.

In February 2007, the Canadian Health Coalition accused the Jean Charest government in Quebec of opening the door to two-tier health care and a major expansion of for-profit surgical clinics.[9] Yet in 2006, only British Columbia was fined by Ottawa for violating the Canada Health Act, although private clinics now operate in six provinces. Quebec now has 16 private MRI clinics, but has not been fined over the past 20 years.

Those who advocate a two-tier medical system never can adequately answer the question as to what the impact will be on our universal health-care system. Where will the doctors, nurses, and technicians that would be needed for private health-care facilities come from?

A group of senior British doctors has urged their Canadian colleagues not to follow the United Kingdom down the road of privatization. For Dr. Danielle Martin, chair of Canadian Doctors for Medicare,

> There is a compelling body of evidence against parallel private insurance in the Canadian context. There is also an inherent conflict of interest for physicians working in dual practice since these physicians could have an interest in promoting longer wait times in the public system to increase use of the more lucrative private system.

Why is the CMA proposing dual practice and private insurance when this would pull nurses, technicians and other needed resources out of the public system?[10]

Good question. Both the Kirby Senate Report and the Romanow Commission had come out in 2002 saying a two-tier medical system would have a detrimental effect on medicare.

Let's look at some more numbers to compare costs. In December 2006, the Canadian Institute for Health Information (CIHI) said that per-capita health-care spending in 2006 in current Canadian dollars was expected to be $4,548, for a total of $148-billion, or 10.3 percent of GDP, with a continuation of the approximate 70-percent-public and 30-percent-private ratio, or $104-billion to $44-billion in 2006.

In contrast to the "runaway" increases often proclaimed by some of the media and conservative health-care critics, total health expenditures in Canada as a percentage of GDP were 10 percent in 1992, and 14 years later, in 2006, only 10.3 percent. In the United States, it's now almost 16 percent and increasing, while per-capita health-care spending is now headed for $7,000 (U.S.). For many years, the United States has spent far more on health as a percentage of GDP than any other country. At this writing, their spending is forecast to reach 20 percent by 2016, while Canada, in a list of the top 30 health spenders, is in eighth place. Some of the countries that traditionally spend more than Canada are Switzerland (11.3 percent), Germany (11.1 percent), Luxembourg, France, Greece, Iceland, and Norway.

Noted American economists Paul Krugman and Robin Wells, writing in the March 23, 2006, edition of *The New York Review of Books*, put the American situation this way:

> We spend far more on health care than other advanced countries – almost twice as much per capita as France, almost two and a half times as much as Britain. Yet we do considerably worse even than the British on basic measures of health performance, such as life expectancy and infant mortality.

The United Nations reports that "on a per capita basis the United States spends twice the Organization for Economic Co-operation and Development average on health care . . . yet some countries that spend substantially less than the United States have healthier populations." For example,

> Malaysia – a country with an average income one quarter that of the United States – has achieved the same infant mortality rate as the United States.
>
> Over 40 percent of the uninsured (in the U.S.) do not have a regular place to receive medical treatment when they are sick, and more than a third say that they or someone in their family went without needed medical care, including recommended treatment or prescription drugs in the last year because of cost.
>
> The uninsured, once in hospital, receive fewer services and are more likely to die than are insured patients. Being born into an uninsured household increases the probability of death before age 1 by about 50 percent.[11]

In terms of public as opposed to private expenditure as a percentage of all spending on health, all the following OECD countries have a higher proportion of public expenditure than does Canada: Austria, Belgium, the Czech Republic, Denmark, Finland, France, Germany, Hungary, Iceland, Portugal, Luxembourg, New Zealand, Norway, Poland, the Slovak Republic, Sweden, and the United Kingdom.

As indicated above, about 30 percent of all health spending in Canada is private. In 2006, public-sector financing of health expenditures came to 70.3 percent of total spending, while private funding made up 29.7 percent. The highest percentage of private spending is, of course, in the United States, at a huge 56 percent.

Total per-capita health spending in the United States is almost two and a half times the OECD average. In Canada, it is one and a quarter times the OECD average.[12] Iceland, Luxembourg, Norway, Switzerland

and the Unites States have higher per-capita total spending than Canada.

If you look at total public-health spending in the OECD and include long-term care expenditures, Iceland, Sweden, Denmark, France, and Germany had higher expenditures in 2005 as a percentage of GDP than did Canada, and 24 countries spent less. The Canadian Institute for Health Information estimates that our total health expenditure, in current dollars, was $131.4-billion in 2004, $139.8-billion in 2005, and $148-billion in 2006, and that total real growth after adjusting for inflation was around 3.7 percent in 2005 and 2006.

Between 1990 and 2002, Germany's public health-care spending was the highest in the OECD, at 8.6 percent of GDP. Canada, at 6.7 percent of GDP, was in eighth place in public health-care spending, behind Germany, Iceland, France, Sweden, Denmark, Norway, and the Czech Republic.

During these years, in terms of private health-care spending as a percentage of GDP, the United States, at 8.1 percent, was far ahead of all other OECD countries. But Canada was up in sixth place as one of the top private spenders, behind only the United States, Switzerland, Greece, New Zealand, and Mexico. Or, to put it another way, 21 countries have higher public health-care funding as a percentage of GDP than Canada, and 23 have a lower share of private funding than Canada. While Canada is at about 30 percent, 16 of the OECD countries have less than 5 percent of their health-care expenditures privately funded.

Looking at total health expenditures as a percentage of GDP, as indicated earlier the United States spends far more than any other OECD country, almost 16 percent. Switzerland in 2003 was next, at 11.5 percent, followed by Germany at 11.1 percent, Iceland at 10.5 percent, Canada and Norway at 10.3 percent, and France at 10.1 percent. The OECD average in 2003 was 8.8 percent. Some of the countries below the OECD average were Italy, New Zealand, Japan, Spain, the United Kingdom, Austria, Finland, and Ireland. Korea, at only 5.6 percent, was at the bottom of the list.

Only the United States, Norway, Switzerland, Luxembourg, and Iceland have higher per-capita health spending than Canada, while 24

OECD countries have lower spending, including countries such as Japan and Korea, which nevertheless have so many impressive health indicators.

LIFE EXPECTANCY AT BIRTH

This is one of the most frequently quoted indicators when comparative health figures are measured. And it's very revealing.

In 1851, the average life expectancy in Canada was only 42.9 years. By 1951, that had increased to 68.7 years, and by 2004 it was 80.2 years. In terms of life expectancy at birth for the period 2005 to 2010, Canada is expected to be in eighth place in the world, at 80.7 years. Japan is forecast to be tops, at 82.8 years, followed by Hong Kong at 82.2, Iceland at 81.4, Switzerland at 81.1, Australia at 81.0, and Sweden at 80.8. The United States is well down the list in 29th place, at 77.9 years. So the average Canadian can be expected to live 2.8 years longer than the average American.

In a list of 50 developed countries, the populations of 42 are expected to have shorter average life spans than that of Canadians, including such countries as Norway, Spain, France, New Zealand, Belgium, Finland, Germany, the Netherlands, the United Kingdom, Ireland, and Denmark.[13] (In OECD countries, there is a "gender gap," whereby women on average live almost six years longer than men.)

Average life expectancy in the least developed countries is only 52 years; in sub-Saharan Africa it is just over 46 years.

In Harlem, New York City, the life-expectancy rate is lower than it is in Bangladesh. Greece, with half of the U.S. per-capita GDP, has a longer life expectancy rate than the United States. While there are several hotly debated explanations for this, it's interesting to note that so many of the countries with much longer lifespans than those of Americans spend a great deal less than the United States does on health care as a percentage of their GDP.

The Economic Policy Institute, based in Washington, D.C., points out that Ireland, Austria, and Finland spend about half as much as does the United States on health care as a percentage of GDP, yet cover 99 to

100 percent of their respective populations with health insurance.
OECD figures indicate that life expectancy in the United States is lower
than it is in 22 OECD countries. Moreover, in the United States, a
steadily rising number of men, women, and children (currently 47
million) have no health-care insurance and many millions more have
totally inadequate insurance. In the words of *The Economist*:

> That so many people should be without medical coverage in
> the world's richest country is a disgrace. It blights the lives of
> the uninsured, who suffer by being unable to get access to
> affordable treatment at an early stage. And it casts a shadow
> of fear well beyond, to America's middle classes who worry
> about losing not just their jobs but also their health-care ben-
> efits. It is also grossly inefficient.[14]

INFANT MORTALITY

Measurements of this key health indicator are based on the number of
deaths of children under one year of age per 1,000 live births. In 1983,
Canada's infant mortality rate was 8.5 per 1,000 live births. By 1992, it was
down to 6.1, and by 2003 it was 5.3. However, the Canadian record is still
disgraceful. Only four of a UN list of 25 OECD countries have a worse
record than Canada.[15] Only one has a record worse than that of the
United States (7.0 per 1,000 live births). Not on the list, the very worst
numbers come from the poor countries such as Turkey, a terrible 29.0 per
1,000 live births, and Mexico at 20.1. That affluent countries such as the
United States and Canada should have such high infant mortality rates
is shameful.

The OECD comments on infant mortality are worth noting:

> The fact that some countries with a high level of health
> expenditures, such as the United States, do not necessarily
> exhibit low levels of infant mortality has led to the conclusion
> that more health spending is not necessarily required to

obtain better results. A whole body of research suggests that many factors outside of the quality and efficiency of the health system, such as income inequality, the social environment, and the individual lifestyles and attitudes are all factors influencing infant mortality rates.

Around two-thirds of the deaths that occur during the first year of life are neonatal deaths (i.e., during the first four weeks).

The lowest infant mortality rates are in Iceland (2.4 per 1,000 live births), Japan (3.0), Finland (3.1), and Sweden (3.1). All the following have rates less than five per 1,000 live births: Norway, the Czech Republic, France, Portugal, Spain, Germany, Belgium, Italy, Switzerland, Denmark, Austria, Australia, Greece, the Netherlands, and Luxembourg.[16]

That Canada has a much higher GDP per capita than many of these countries and at the same time a much worse infant mortality rate should be of great concern. Several leading Canadian health-care authorities attribute Canada's poor showing to our relatively large aboriginal population compared to the aboriginal population in most other OECD nations.

While the gap has narrowed as overall aboriginal health has improved, infant mortality among aboriginal peoples in Canada is still about one and a half times the non-aboriginal rate, while the aboriginal birth rate continues to be much higher. Setting aside the high infant mortality rate among aboriginal peoples allows Canada's record to compare favourably with the top OECD countries, but our treatment of our aboriginal people can only be described as disgraceful.

Projections for infant mortality rates in the period from 2005 to 2010 place Canada down in 19th place, at 4.8 deaths per 1,000 live births. The lowest projected rates include Singapore at 3.0, Iceland and Japan at 3.1, Sweden at 3.2, Norway at 3.3, South Korea at 3.6, Finland and Hong Kong at 3.7. Put another way, 18 OECD countries are expected to have a lower infant mortality rate than Canada.

Recent comparisons indicate that if the United States had an infant mortality rate as low as Cuba's, an additional 2,200 American babies would be saved every year. If the American rate was as good as Singapore's, 18,900 babies a year would be saved.

Just over 0.5 percent of U.S. newborns die in their first month. Of all industrialized nations, only Latvia has a higher rate.

UNDER-FIVE MORTALITY AND LOW-BIRTH-WEIGHT INFANTS

In 2004, all the following countries had lower under-five mortality rates than Canada: Austria, Belgium, Cyprus, Denmark, France, Germany, Greece, Italy, Liechtenstein, Monaco, Portugal, Spain, Switzerland, the Czech Republic, Finland, Japan, Norway, San Marino, Slovenia, Sweden, Iceland, and Singapore.[17] Which puts Canada way down in 23rd place.

The number of low-birth-weight infants is another key standard-of-living indicator. In the period from 1998 to 2004, 21 countries had a better record than Canada in this respect. Canada had low-birth-weight infants in the 6 percent range. All of the following were at or below 4 percent: Iceland, Sweden, Finland, Korea, Estonia, Lithuania, Bosnia and Herzegovina, Albania, and Western Samoa.[18] Note once again that many of these countries had lower or much lower GDP per-capita figures than Canada.

The United States had about 7.9 percent of all births classified as low-birth-weight infants.

MORTALITY PER 100,000 POPULATION, 2002

The OECD explains that "mortality rates are, paradoxically, the most common measures of a population's health, since mortality statistics remain the most widely available and comparable source of information on health problems. Age standardizing death rates remove the effects of variations in the age structure of populations across countries and over time."

The leading causes of death in OECD countries are related to cardio-vascular diseases, cancer, and diseases of the respiratory system. In a list of 20 OECD countries, Canada has the sixth highest fatality rate in the period 30 days after a heart attack and the eighth highest rate of breast cancer mortality.[19]

Twenty OECD countries have death rates higher than Canada's, which is 565 per 100,000. Six have lower rates: Japan, Australia, Switzerland, Iceland, Italy, and Spain, whose rate is almost identical to Canada's. The lowest mortality rate, at only 449 per 100,000, belongs to Japan, which also tops the life-expectancy list.

Canada has one of the three lowest cerebrovascular mortality rates, but somewhat higher than OECD average lung cancer and breast cancer rates, and a slightly better prostate cancer mortality rate.

Road accidents killed over 120,000 people in OECD countries in 2002. Canada was in 18th place in fatalities per million population, somewhat below the OECD average, as we were in suicides. The highest suicide rates were in Korea, Japan, Finland, and Hungary. The lowest were in Greece, Italy, and the United Kingdom.[20]

The United Nations estimates that some 56,000 children and adults in Canada were living with HIV at the end of 2003. The adult prevalence rate (ages 15 to 49) was 0.3 percent. Sixty-three countries have lower rates. The U.S. rate of 0.6 percent was double Canada's.

Some countries have horrific HIV rates. Botswana's rate was 37.3 percent, Lesotho's 28.9 percent, Namibia's 21.3 percent, South Africa's 21.5 percent, Swaziland's 38.8 percent, Zimbabwe's 24.6 percent, and four other countries, all in Africa, have rates over 10 percent. In 2003, there were some 1,000,000 children and adults living with HIV in the Congo, 1,500,000 in Ethiopia, 1,200,000 in Kenya, 1,300,000 in Mozambique, 3,600,000 in Nigeria, a staggering 5,300,000 in South Africa, 1,600,000 in Tanzania, and 1,800,000 in Zimbabwe. There are about 1,000,000 adults and children living with HIV in the United States.[21]

In 2006, the Public Health Agency of Canada said that almost 15,000 Canadians had HIV/AIDS but didn't know they did, while some 58,000 other people in Canada were living with the virus at the end of the year, about 80 percent of them men.

OBESITY

In the summer of 2006, Statistics Canada said that the Canadian national average for obesity was a shocking 23 percent. About the same time, a University of North Carolina study said that for the first time there were more overweight people in the world than the number that were undernourished, and that while the number of hungry people was falling slowly, the number of those who were obese was increasing "at an alarming rate."

Britain and the United States have two of the highest rates of obesity, Japan one of the lowest.

Statistics Canada has also reported that in 2004, 1.6 million young people aged two to 17 were overweight, an increase from 15 percent in 1989 to 26 percent. Of those, 507,000 were considered obese.

In international comparisons, Canada does not do well, with the fourth highest obesity rate in an OECD list of 30 countries. All the following countries have obesity rates less than half of Canada's: Japan, Korea, Switzerland, Norway, Italy, Austria, Denmark, France, and Sweden. (A recent Statistics Canada study released in June 2006, but using different standards than the OECD, puts the too-heavy figure for Canada at 24.3 percent.) If Canada is bad, however, the United States, where the rate of obesity has more than doubled over the past 15 years, is appalling. Just under 30 percent of Americans over the age of 15 are classified as obese.

The OECD puts it bluntly: "In many OECD countries, the growth in overweight and obesity rates in children and adults is rapidly becoming a major health concern." Obesity often brings with it health problems such as hypertension, diabetes, high cholesterol, cardiovascular diseases, respiratory problems, and some forms of cancer.

Over half the adults in the United States, Mexico, the United Kingdom, Australia, the Slovak Republic, Greece, New Zealand, Hungary, Luxembourg, and the Czech Republic are now classed as either overweight or obese.

CORONARY PROCEDURES

For coronary bypass procedures, the figure for the United States of 161 per 100,000 population compares to 98 for Canada and only 70 for the OECD average. For coronary angioplasty procedures, the United States is at 426 per 100,000 population, Canada is at only 140, while the OECD average is 150.

There is much debate as to why the U.S. figures are so high. Switzerland, France, Spain, and Italy have numbers ranging from only 19 to 46 per 100,000 for bypasses, for example. Many suggest that the profit motivation for U.S. doctors and hospitals is a major factor encouraging operations that may, in fact, not be considered necessary in other countries. In the United States, there tends to be more of a "sell your product" health-care mentality, whereas in Canada and other countries with more public health care, there tends to be more caution about using unnecessary procedures to avoid budgetary problems. Nonetheless, ischemic heart disease mortality rates are much lower in countries such as Australia and Canada than they are in the United States, despite the fact that the United States has by far the highest overall rate of coronary re-vascularization procedures of any OECD country (596 per 100,000, compared to 238 in Canada and 212 in Australia).[22]

Obviously, the high rates of obesity in the United States contribute to greater heart and other problems.

ACUTE-CARE BEDS

Twenty OECD countries do better than Canada in the number of acute-care beds (in Canada, 3.2 per 1,000 population) while only nine countries have fewer beds, including the United States, at only 2.9. Topping the list is Japan, at 8.9 beds, followed by Germany at 6.6, and Austria and the Czech and Slovak Republics are all at over six acute-care beds per 1,000 population. Canada is below the OECD average of 4.2.[23]

Canada does poorly compared to other developed countries in MRIs and CT scanners: 16th of 27 countries in MRI units and 19th in CT scanners. Canada has only 4.5 MRI units per 1,000,000 population, compared to the OECD average of 7.2, and we have only 10.3 CT scanners, compared to the OECD average of 17.6. Japan, at 35.3 MRI units and 92.6 CT scanners, is well ahead of all other OECD countries. Mexico and Poland are at the bottom of both lists.[24] In one study, 51 percent of Canadian physicians reported patients facing long waits for diagnostic tests, compared to only 6 percent in Australia.

Can we afford more MRIs? Or perhaps a better question is to ask how is it that we have done so poorly compared to other countries in acquiring such vitally important diagnostic aids? For some answers, see the chapter on taxes in this book.

EXPENDITURES ON PHARMACEUTICALS

Spending on pharmaceuticals has risen dramatically in recent years in OECD countries. Only in Iceland, Greece, Luxembourg, the Czech Republic, and Japan has it declined as a share of total health expenditures (from 1997 to 2007). During these years, pharmaceutical spending in Canada in real terms grew at an annual rate of 6.9 percent. In comparison, in the United States during the same period, pharmaceutical spending grew by an average of 9.5 percent per year. The OECD comparative figure was 5.6 percent.

Among OECD nations, Canada is the third highest per-capita pharmaceutical spender and ninth highest when such spending is measured as a percentage of total health spending. Twenty years ago, drug costs made up about 9.5 percent of our health-care costs. Today, it's about 17 percent. Bulk buying, as is done by countries such as New Zealand, would help reduce our current $25-billion drug costs substantially, but our politicians can't seem to move on what would certainly be an important cost-saving program. In Canada, total personal expenditures on medical and health services amounted to $40.68-billion in 2005, of which just over $15-billion was for drugs and pharmaceutical products. The Canadian Institute for Health Information says,

The category of drugs ranks second after hospitals in terms of its share of total health expenditures. In 1997, expenditure on drugs overtook spending on physician services. The share of total spending accounted for by drugs grew from a low of 8.4 percent in the late 1970s, to 16.6 percent in 2004. In 2007, drugs are ranked second with a share of 17 percent of total health expenditures.

Overall, where do public health-care dollars go in Canada? In 2006, 30 percent went to hospitals and 9.6 percent went to other health-care institutions. Some 13 percent went to physicians, 11 percent to other health-care professionals, and about 15.9 percent went for capital costs, public health costs, and administration. The share spent on health research was 1.6 percent.[25]

SMOKERS

The World Health Organization says that tobacco is the second major cause of death in the world and is directly responsible for about one in ten adult deaths worldwide, or about five million deaths each year.

Canada has the lowest percentage of OECD adults smoking tobacco daily, some 17 percent, followed by the United States and Sweden at 18 percent. Ten other countries are below the OECD average of 26 percent. The countries with the worst records are Greece, at 35 percent, Hungary, 34 percent, Luxembourg, 33 percent, and 12 other countries, including Bulgaria, Japan, Spain, Russian, and Ukraine.[26]

In many countries, there is a large gender difference. For example, the rate of smokers is 63 percent of men in Korea and only 5 percent of women. In Canada the percentage of men who smoke fell from just under 44 percent in 1981 to 22 percent in 2005, and for women from 32 percent down to 16 percent in 2005 during the same years. Only 1 percent of the women in Cuba smoke and only 17 percent of the men. In Japan, only 15 percent of the women smoke, but 47 percent of the men do. In the United States, 19 percent of the women still smoke and 24 percent of the men.

Canada has the highest percentage of people in the industrialized world using marijuana on a regular basis, more than four times the global rate. Of Canadians aged 15 to 64, 16.8 percent smoked marijuana or used other cannabis products in 2004. The number of Canadians using marijuana doubled over the past decade.[27]

ALCOHOL CONSUMPTION IN LITRES PER CAPITA

The OECD 2003 average for individual alcohol consumption was 9.6 litres. Eight OECD countries had a lower average consumption in 2003 than did Canada (at 7.8 litres). Turkey was the lowest, at only 1.5 litres, followed by Mexico at 4.6 litres, Norway at 6.0, Iceland at 6.5, Sweden at 7.0, the Slovak Republic and Japan at 7.6. The level in the United States was 8.3 litres. The highest alcohol consumption was in Luxembourg, at 15.5 litres. France's average consumption was 14.8 litres, Ireland's was 13.5 litres, and Hungary's was 13.4 litres. Eleven other countries were above the OECD average.

In most OECD countries, there has been a reduction in alcohol consumption and in deaths from liver cirrhosis. According to the OECD, wine consumption has been increasing in many traditionally beer drinking countries. Alcohol consumption in Italy and France has dropped substantially, but it has increased substantially in Iceland and Ireland.[28]

Finally, with all of our system's problems, a report confirms the advantages of Canada's health-care system over the American system. In April 2007, a report in *Open Medicine*, a new online Canadian medical journal, said,

> Canada's much-maligned health system produces as good or better outcomes as the vaunted U.S. system, and it does so at less than half the cost.
>
> A team of 17 Canadian and U.S. health-care researchers

crunched data from 38 existing studies from both countries published between 1955 and 2003.

Canadian patients had at least as good an outcome as their American counterparts, if not better.

The authors said that "the fundamental message of this study is that the solutions to Canada's health-care problems lie not in resorting to U.S. style private funding or for-profit delivery, but rather in strengthening publicly funded health care delivered by not-for-profit providers."[29]

Regarding the ongoing pressure for a two-tier health-care system in Canada, here are the words of Dr. Arnold S. Relman, Harvard emeritus professor of medicine, and emeritus editor-in-chief of the *New England Journal of Medicine,* who favours the elimination of all for-profit facilities: "The facts are that no one has ever shown, in fair, accurate comparisons, that for-profit makes for greater efficiency or better quality, and certainly have never shown that it serves the public interest any better. Never."[30]

Despite some recent improvements, there are still major problems in Canada's health-care system. There remain shortages of medical beds and nurses. Only one in five family physicians are accepting new patients. The number of physicians in Canada is still far below OECD average levels and surveys indicate that more than 4,000 Canadian doctors plan to stop practising within the next two years.[31]

All of this said, a January 2008 report by a British researcher in *Health Affairs* says that "Canadians are getting excellent value for the money" compared to the citizens of 18 other countries studied, and the outcomes are substantially better than those for their southern neighbour.

For readers who want more detailed health-care information, the excellent Canadian Institute for Health Information (www.cihi.ca) has a long list of valuable reports that can be downloaded free of charge in both English and French.

POVERTY IN CANADA

"We're not the lovely people we think we are."

"Like slavery and apartheid, poverty is not natural. It is man-made and it can be overcome and eradicated by the actions of human beings."
— NELSON MANDELA, BBC, FEBRUARY 5, 2005

Year after year in Canada, the public opinion polls are clear. After health care, and more recently after the environment, but well ahead of lower income taxes or reducing government debt or increasing military/defence spending, Canadians place a constant and a very high priority on reducing child poverty. But of course you don't reduce child poverty without lowering the level of overall poverty.

Let's look at both problems and see how Canada measures up. First, a reminder. In 1989, the House of Commons passed their now notorious all-party unanimous resolution promising to wipe out child poverty in Canada by the year 2000. At that time, using the before-tax system of measuring poverty (see Appendix Two for an explanation as to how poverty is measured), 15.1 percent of children in this country were living in poverty. By 2004, despite substantial economic growth and huge wealth creation, the number of poor children had grown to 17.7 percent, or almost 1.2 million children. Statistics Canada also said that 12.5 percent of all Canadian families, 34.5 percent of immigrants who had been in Canada less than 10 years, and almost 50 percent of female lone parents were classified as living in "low-income" situations.

At the end of March 2006, the *Globe and Mail* ran a front-page headline saying, "Growth Spurs Decline in Poverty." The following month, the paper had a glowing editorial applauding Canada's success in our war against poverty. Using *very* conservative poverty measures, well below those used by the United Nations, the OECD, and the European Union, the *Globe* applauded the "heartening" facts that in 2004 only 3.5 million Canadians lived in poverty, that only 14 percent of those who were employed had full-time jobs paying less than $10 an hour, and that only 865,000 Canadian children were living in poverty. For the *Globe,* "All in all, it is a comforting picture."

Let's take this so-called "comforting picture" and compare it with the pictures presented by the OECD, by Unicef, and by other international organizations and experts measuring poverty.

Here are conservative figures for the percentage of countries' populations living below the poverty line for the period from 1990 to 2000, from the 2003 United Nations *Human Development Report.*

Slovakia	2.1%	Denmark	9.2%
Luxembourg	3.9%	Switzerland	9.3%
Czech Republic	4.9%	Spain	10.1%
Finland	5.4%	Austria	10.6%
Sweden	6.6%	Japan	11.8%
Hungary	6.7%	Ireland	12.3%
Norway	6.9%	Estonia	12.3%
Germany	7.5%	United Kingdom	12.5%
Belgium	8.0%	Canada	12.8%
France	8.0%	Israel	13.5%
Netherlands	8.1%	Italy	14.2%
Slovenia	8.2%	Australia	14.3%
Poland	8.6%	United States	17.0%

So, compassionate Canada is way down in 22nd place.

The Canadian Council on Social Development commented:

In 1989, Canada made a commitment to end child poverty. Instead, successive governments have created ever more elaborate ways to measure it. With the new Market Basket Measure, we now have six different poverty measures which all show the same thing: no matter how you count them, there are too many poor people in Canada.

In February 2007, an updated Unicef report said that in a list of 25 OECD countries, 18 had lower rates of child poverty than Canada, whose rate was almost 15 percent. The following countries had child poverty rates ranging between 2.4 percent and 4.2 percent, *all less than a third of the rate in Canada:* Denmark, Finland, Norway, and Sweden. All of the following had rates of under 10 percent: Switzerland, the Czech Republic, France, Belgium, Hungary, Luxembourg, and the Netherlands.

How's that for Canada, a country with the world's eighth highest GDP per capita?

At the bottom of the list, with the worst child poverty rates? Mexico, at about 27.7 percent. And next worst? Can you guess? The United States, of course, at about 22.7 percent. Year after year, the United States, the world's wealthiest country, is right down near the very bottom of the barrel.

In another comparison, this time using the most conservative poverty rates, the 2005 OECD study of social indicators, *Society at a Glance,* said that back in 2000 Canada was way down in 18th place among developed countries in the percentage of its population living in poverty. Canada's rate of 10.3 percent was more than double the rate for Denmark, at 4.3 percent, and the Czech Republic, at 4.4 percent. All of the following were between 5 and 7 percent: Luxembourg, Finland, Sweden, the Netherlands, Norway, France, and Switzerland.

The OECD's "poverty gap" is a measurement of the difference between the average income of the poor and the 50-percent-of-median-income threshold. Here, too, Canada does poorly. Seventeen countries have a smaller poverty gap, and Canada's is higher than the OECD average.

More recently still, in an updated list of 27 developed countries in the 2006 United Nations *Human Development Report,* 16 countries had a lower percentage of their populations living below the United Nations' broadly accepted poverty line when it was computed for the years 1994 to 2002. And shockingly, in the most recent UN calculations, Canada is back down at 19th in a list of 26 OECD countries.

So what do you think? How is it that so many countries have child poverty rates only a quarter of Canada's, or a third, or less than half? It's no secret. It's not magic. Through decent social programs, Denmark has been able to reduce its child poverty rate from 11.8 to 2.4 percent, Finland from 18.1 to 2.8 percent, Norway from 15.5 to 3.4 percent, Sweden from 18 to 4.2 percent, the Czech Republic from 15.8 to 6.8 percent, France from 27.7 to 7.5 percent, Belgium from 16.7 to 7.7 percent, Hungary from 23.2 to 8.8 percent. And there are many other examples of public policy sharply reducing rates of child poverty.

So what's wrong with wealthy Canada? See the chapter that follows on comparisons of international rates of social spending.

Timothy Smeeding of Syracuse University is the director of the renowned Luxembourg Income Study. In his October 2005 *Poor People in Rich Countries* (www.lisproject.org), he compares 11 countries: Canada, the United States, Ireland, the United Kingdom, Austria, Belgium, Germany, the Netherlands, Italy, Finland, and Sweden. Among his major points:

A majority of cross-national studies define the poverty threshold as one-half of national median income. . . .

Alternatively, the United Kingdom and the European Union have selected a poverty rate of 60 percent of the median income. . . .

The United States makes the least anti-poverty effort of any nation, reducing relative poverty created by market incomes by 28 percent compared to the average reduction of 61 percent. . . .

In most rich countries, the relative child poverty rate is 10 percent or less; in the United States it is 21.9 percent. What seems most distinctive about the American poor, especially poor American single parents, is that they work more hours than do the residents of other nations, while also receiving less in transfer benefits than in other countries. . . .

In the 11 rich countries in Smeeding's study, six have lower overall poverty and child poverty rates than Canada: Germany, Belgium, Austria, the Netherlands, Sweden, and Finland.

As a result of government measures to reduce poverty, Canada has been able to cut the overall rate by 46 percent. But eight of the 11 countries in Smeeding's study have been more successful, some of them much more successful. Finland, for example, reduced its overall poverty by 81 percent, and Sweden and Belgium by 77 percent. Only the United States and Italy have a record worse than Canada's in reducing poverty rates in single parent households.

It's interesting to use the Luxembourg Income Study to compare child poverty in the three North American free-trade agreement (NAFTA) nations. In Mexico, since 1986, it's never been below 23 percent. In the United States, the lowest it's been is about 22.5 percent. Using similar measurements, during the same years in Canada, it has hovered around 15 percent.

In November 2006, the widely based Campaign 2000[1] coalition produced its annual report, which began:

> The rate of child and family poverty in Canada has been stalled at 17–18% over the past 5 years despite strong economic growth and low unemployment. In fact, data from Statistics Canada show that over the past 25 years Canada's child poverty rate has never dropped below the 15% level of 1989 when the House of Commons resolved to end child poverty.

The report shows that 1,196,000 children, almost one in every six, were living in poverty, that about one in three poor children live in families where at least one parent was working full-time, and that a great many poor families in Canada are very poor. For example, "The average poor female lone parent would need $9,400 a year additional income just to bring them up to the poverty line," and average low-income families would need an extra $7,200 a year just to be able to reach the Statistics Canada low-income line.

The report also notes that "even with a booming economy, Alberta's child poverty rate is double digit and has fluctuated between 14% and 15% since 1999. The child poverty rate in British Columbia is 23.5%."[2] How's that for progress?

Want yet another source? In the spring of 2006, Statistics Canada reported that by their most conservative after-tax measures, 12.8 percent of Canadian children lived in low-income families. Bear in mind that these figures somehow fail to count the terribly poor aboriginal children living on Canada's Indian reserves.

You read the junk in the *National Post* about poverty in Canada and still want yet another source? In its 2006 Economic Survey of Canada, the OECD has a chart of poverty rates for jobless households in 23 countries. The highest poverty rate was for the United States, but Canada's was the second highest.

And yet another? In 2006, the United Nations Committee on Economic, Social and Cultural Rights once again sharply criticized Canada, as it did in 1998. The criticism was aimed at our high rate of poverty, our totally inadequate unemployment and welfare benefits, our far too low minimum wages, our treatment of our aboriginal peoples, the fact that social assistance benefits are lower now than they were a decade ago, the "discriminatory" child benefit clawback, and Canada's failure to respond to similar UN criticisms from eight years earlier.

As far as the *National Post* is concerned, let's turn briefly to its founder. In the 1990s, the colossally arrogant Conrad Black, owner of dozens of Canadian daily newspapers, incredibly told us that "caring and compassion really means socialism." In a speech in Edmonton in 1992, Black complained about Canada's "extravagant welfare programs which cause money to be skimmed to people who haven't earned it." Speaking of skimming money . . .

So what has happened to compassion in Canada? Do we care about poor children or not? We know from numerous studies that poor children tend to have far more disabilities and poorer functioning levels of vision, speech, and mobility, shorter lifespans and more chronic illness, as well as engage in more criminal activity and receive harsher treatment in the justice system, and many other problems.

And we know from many other reliable studies that apart from the misery and despair that comes with poverty, it makes no economic sense at all. It's far less costly to tackle poverty than it is to pay for its consequences. Ed Finn, writing in the June 2007 issue of *The CCPA Monitor*, points out that

> the Center for American Progress (CAP), a progressive think-tank in Washington, recently did a study on the economic costs of child poverty in the United States. Their researchers' estimated figures are staggering. They calculated that Americans who were poor as children – and there are now 37 million of them – are much more likely than other citizens to commit crimes, to need more health care, and to be less productive in the workforce.
>
> One CAP researcher, Harry J. Holzer, described the results of their study to a House Ways and Means Committee hearing last January. He told the stunned Congressmen that the costs to the U.S. in crime, health care, and reduced productivity associated with childhood poverty amount to an estimated $500 billion a year.

What, then, can we say about our own politicians? Over and over again, Jean Chrétien talked about building a fairer society where no one got left behind, but every year the gap between rich and poor increased, with the rich getting much richer and the poor staying poor. On the issue of poverty in Canada, Paul Martin was one of the greatest hypocrites in the modern history of politics in this country. And Stephen Harper? Has he even mentioned the topic in a speech?

What does it say about Canadians and our political leaders when for decades we've had growing homelessness and a mushrooming of food banks across the country? And what does it say about us when our welfare rates for poor children and the disabled don't come anywhere near the poverty line? And what does it say when, in the face of these facts, tax cuts for big corporations and the already well-to-do are top government priorities, leading to greater corporate concentration of wealth and much greater inequality of incomes?

And what does it say about the media in Canada when the issue of child poverty is either attacked, in the most strident terms, as being exaggerated or else receives so little attention? Carol Goar of the *Toronto Star* wrote about a press conference organized by the late June Callwood and three other children's activists to draw attention to the issue prior to the 2006 federal election.

> Only four journalists showed up, two from religious publications, one from Omni television and one from the *Star*.
>
> The pile of press kits sat pathetically on a table. The child-care and church leaders who gathered for the event tried to hide their disappointment.
>
> Callwood asked how countries such as Hungary and Poland – which ranked 25th and 28th on the global wealth scale – can afford to treat their children better than Canada.
>
> Callwood's colleague, Rabbi Arthur Bielfeld, said, "I believe in the decency of Canadian society, but I'm becoming increasingly restive. We are not responding to the despair around us. We're not the lovely people we think we are."[3]

Poverty is not often mentioned in federal or provincial elections, in leadership contests, or in parliament. Even at the 2006 NDP national convention in Quebec City, the issue was almost invisible, as it was later in the year at the Liberal leadership convention in Montreal.

Have you ever visited a food bank in Canada? Or, better still, worked in one for even a day? It's too bad that our extreme-right-wing poverty deniers don't have time to do so. They'd find that about 40 percent of the over 720,000 Canadians who were forced to rely on food banks in 2007 were children, and that despite our relatively affluent society, food bank usage has increased by some 20 percent over the last five years, and the number of children relying on them has almost doubled since 1989.

Many of those who have to rely on food banks are working men and women whose income just isn't anywhere near enough to look after their families. Many are poor single mothers. Many are disabled. Others are poor students. In their November 2006 *Hunger Count* report, the Canadian Association of Food Banks said that a major percentage of their clients were people who report that they are not able to get more than 25 hours of work in a week.

Food bank use in Canada has increased by almost 80 percent during the past decade. Increasingly, low-paid working poor and their children are using food banks. Why? In Toronto, people who use food banks are spending on average 73 percent of their total income on rent. Somehow only 22 percent of Toronto's unemployed qualify for employment insurance benefits. And 70,000 people in the city are on waiting lists for affordable housing.

The average two-bedroom apartment in Toronto rents for $1,052 a month. A single mother with two school-age children gets a grand total of $1,184 a month in social assistance. After paying an average rent, that leaves $132 a month for food, for clothing, for school supplies, for utilities, for kitchen and bathroom supplies, for recreational activities, etc., etc. That's $132 per month for a family of three. The average poor family in Toronto is almost $10,000 short of adequate money for rent, food, and public transit, and the fastest-growing group using shelters is children.

Brian Mulroney's Conservative government cut a massive $2-billion from federal housing programs. The Chrétien government did nothing to restore public housing and downloaded responsibility for it to the provinces. Then Paul Martin abolished the housing ministry.

In Britain, Prime Minister Gordon Brown has announced a remarkable $17-billion affordable-housing plan that will produce 50,000 units of social housing a year for the next three years, with a goal of three million new affordable homes by 2020. By contrast, a *Toronto Star* editorial correctly says that

> In Canada there is no national housing strategy. Worse, there is a lack of political will to develop one, despite a growing homelessness crisis and huge waiting lists for subsidized housing across the country.
>
> Canada is alone among the major industrialized countries in not having a national housing strategy. Only 5 percent of the housing stock in this country is social housing, one of the lowest levels in the world.[4]

There are now an estimated 1,500 homeless in Vancouver, but if no additional low-cost housing is built, that number will likely increase to over 3,000 by 2010. One new estimate puts the number of homeless in British Columbia at well over 10,000. And while Ontario's Liberal government under Dalton McGuinty promised 20,000 affordable housing units for that province, after three years in office only 6 percent of these were built.

In Ontario, over 122,000 households were waiting to get into affordable housing in 2006. The average household income of people on the waiting list is only slightly over $12,000 a year. While Paul Martin spoke frequently about poverty in Canada, if you were to add up all the affordable public housing units built in Canada during all the years that he was minister of finance, the total wouldn't equal even a single year of public housing construction in the 1980s. In 1980, 24,168 affordable residences were built. In 1998 only 550 were built.

For those of us who have been concerned for many years about the very high levels of poverty in Canada, the publication of the 2006 edition of the *Hunger Count* report by the Canadian Association of Food Banks can only be described as dismaying. The first food bank opened in Edmonton in 1981. It was thought to be just a temporary measure. In 1989, there were 159 food banks in Canada. Today, there are food banks in every province and territory, 649 in all. In 1981, those who organized the first food bank worried that while they felt a moral obligation to help the hungry, their actions might lessen the pressure on government to do something about the urgent problem. Unfortunately, their concerns were justified. Since the House of Commons promised to abolish child poverty, food bank usage in the same period has more than doubled.

Incredibly, in an era of unprecedented affluence, over one third of all food banks in Canada report that they have difficulty meeting the demand for food from hungry men, women, and children. An editorial in the *Toronto Star* reports that

> The Foodpath food bank in Mississauga is in desperate need of help as it tries to meet growing demands for its services. That such a crisis exists in a 905 community may come as a surprise to those who believe poverty, homelessness and hunger are problems unique to the city of Toronto. But the Mississauga charity helps 5,500 people every month. Half are children.

Meanwhile, the core operation of the Canadian Association of Food Banks receives zero government funding, and only in Quebec and Nova Scotia is there more than minute government assistance. According to the *Hunger Count* report,

> In one month in 2006, 753,458 Canadians obtained food from a food bank; 41 percent were children. Contrary to popular assumption, many food banks can only provide a few days' worth of food. Food shortages have forced dozens of food banks to cut back on food hampers, turn people away, and even shut down for several days.[5]

Welfare rates in Canada are pathetically inadequate (see the chapter in this book on welfare). As a result, many of those who must resort to food banks are also dependent on welfare.

Greg deGroot-Maggetti, of Citizens for Public Justice, opens his foreword to *Hunger Count* with a famous quote from the late Brazilian bishop Dom Halder Camara: "When I feed the poor, they call me a saint. When I ask why they are poor, they call me a communist." The wonderful Sue Cox, who in December 2005 stepped down after 17 years as the executive director of the Daily Bread Food Bank in Toronto, says, "My greatest regret is that we have failed to put the serious issue of hunger on the public agenda."

One further thought on the subject of poverty in Canada. During the 19 years since the House of Commons resolution promising to abolish child poverty, Canada's GDP more than doubled, increasing by almost $880 billion. But the most recent Statistics Canada figures put our child poverty rate at exactly the same level as it was back in 1989.

(In Appendix Two, you will find a brief discussion of how poverty is measured and some additional comments on poverty in the United States. The National Council of Welfare website, www.ncwcribes.net, has a report, *Solving Poverty: Four Cornerstones of a Workable National Strategy for Canada*, which can be downloaded free of charge.)

PART TWO

ABORIGINAL PEOPLES IN CANADA

THE SHAMEFUL NEGLECT OF APPALLING, DISGRACEFUL, GRINDING POVERTY

"Many of us just don't give a damn."

As we all know, Canada has ranked at or near the top of the United Nations Human Development Index for many years. At the same time, our aboriginal peoples rank 63rd on the same scale.

The average life expectancy rate for Canada's aboriginal people is seven years shorter than the lifespan for non-aboriginal Canadians. The levels of diabetes, disability, suicide, poverty, and unemployment among aboriginals, particularly those living on reserves, are all significantly higher than the levels among non-aboriginals. And the disability rate among First Nations children is over twice the national average.

According to Laurel Rothman of Campaign 2000 and Assembly of First Nations head, Phil Fontaine,

- more than four in ten First Nations children are in need of basic dental care they cannot afford;
- nearly 100 First Nations communities must boil their water;
- mould contaminates almost half of First Nations households;
- diabetes is three to five times more common than the Canadian average and tuberculosis is eight to ten times more common;
- forty percent of aboriginal children whose homes are off-reserve live below the poverty line.[1]

Aboriginal people are about 3.8 percent of Canada's population, but they make up about 20 percent of all prison inmates. The prison rate of 1.6 percent is eight times the rate for other Canadians. There are over 17,000 native inmates in Canada's prisons, where they face systemic discrimination.[2] In Manitoba, more than 68 percent of those sentenced to custody in 2003/2004 were aboriginals although they made up only 10.6 percent of the province's population. In Saskatchewan, it was more than 80 percent with less than 10 percent of the population. In Alberta, it was almost 39 percent with just over 4 percent of the population.

In Toronto, aboriginals make up about 1 percent of the population but over a quarter of the homeless. They also remain homeless much longer than other homeless people, on average about five years.

The unemployment rate for young aboriginal workers is just under 23 percent, about double the rate for all young workers. In 2005, the unemployment rate for off-reserve aboriginals in Western Canada was 2.5 times higher than the rates for non-aboriginals.[3]

Between 1997 and 2000, the average homicide rate for aboriginal people was almost seven times higher than it was for non-aboriginals.[4]

In 2001, a startling 41 percent of aboriginal children under age 15 were poor. And 37 percent between the ages of 15 and 24 were poor. These figures were more than double the rates for non-aboriginal children. Yet somehow people registered under the Indian Act living on reserves are not included when poverty rates are calculated.

Phil Fontaine, speaking about the 30 percent of aboriginals who live on reserves, said bluntly:

> The underlying problem is the impoverished state of First Nations communities.
>
> We exist with poor housing, poor schools, poor access to quality health care, poor drinking water, and the pressure as a result of this grinding poverty is just overwhelming for too many of our people.

Thirty-five percent of on-reserve aboriginals are on welfare.

Michael Mendelson of the Caledon Institute of Social Policy has pointed out that only 16 percent of the general population between the ages of 20 and 24 have not finished high school, but for natives living on reserves in the same age group it's an appalling 58 percent. Mendelson writes, "What do you suppose their young men and women will do with their lives? . . . The only difference between this and the kind of disasters that grab headlines and emergency funds is that it will take longer for the destruction to become obvious."

John Ibbitson of the *Globe and Mail* is right to the point:

> Because we hold neither our leaders nor ourselves to account, because we use such words as "consultation" and "consensus" instead of "crisis" and "emergency," because so many of us just don't give a damn, millions of dollars that should be going to native education are being wasted, leaving a generation of students untaught.
>
> Everyone in the federal government and within the Indian community knows this is happening. But it is easier for politicians to blame each other, and for native leaders to blame the politicians, than for anyone to say: "Stop. This is our responsibility. We have to act."
>
> Each year, Indian and Northern Affairs Canada transfers $9.1-billion to first nation governments. Of that, $1.5-billion, or 16.5 percent, is earmarked for education. How much of that amount is actually spent on schools? Nobody knows.[5]

Ibbitson quotes Peter Garrow, the Assembly of First Nations education director, as saying, "they have to make some tough decisions" on reserves, where education needs must compete with all other needs, such as health and housing.

Does getting an education help? In 2005, aboriginal people who had a university degree had an 84 percent employment rate.[6] But only 4 percent of aboriginals have a university degree.

The historic Kelowna Accord was signed in November 2005 with much applause across the country and almost universal approval from the provinces, the territories, and from our native peoples. The objective was to step up the level of aboriginal health care, housing, education, and general living conditions. There was, however, just a small problem: The Harper Conservative government junked the agreement, abandoning years of negotiations, even though each of the five Conservative premiers had supported the accord, along with the Assembly of First Nations, the Métis, and the Inuit. According to John Ibbitson, "The Conservatives lack the political courage to confront, head on, the overriding social policy challenge of our time: eliminating aboriginal poverty on and off reserve."[7]

Between 1990 and 2001, the suicide rate per 10,000 youths 15 to 24 in Canada was 11.9. In the United States, it was 9.9. In Mexico, it was only 4.7. Why was it so high in Canada? The suicide rate for aboriginal youth in this country is three times that of non-aboriginal youth.

But then again, it seems many of us just don't give a damn.

CANADIAN SOCIAL POLICY

COMPARED TO MOST OTHER DEVELOPED COUNTRIES, WE ARE A DISGRACE, OUR POLICY "AN UTTER DISASTER"

"Appalling" is the only way to describe Canada's social spending to assist children, low-income households, the needy elderly, the sick and disabled, the unemployed, and other disadvantaged persons who need help. Compared to most other developed countries, we are a disgrace. The National Council of Welfare has described social policy in Canada as "an utter disaster."

When all federal, provincial, and municipal social spending is added up, Canada is way down in 25th place among the 30 OECD countries in terms of social spending as a percentage of GDP. (The Unites States, the world's wealthiest nation, is in 26th place, beating out only Ireland, Turkey, Mexico, and Korea).[1]

In May 2007, Save the Children also put Canada down in 25th place in terms of how our society treats children, and we're at the very bottom of another list of OECD nations when it comes to our investing in early learning.

Heading the list in overall social spending as a percentage of GDP is Sweden, at just over 31 percent, followed by France at 28.7 percent, and Denmark and Germany at 27.6 percent. Canada, the world's ninth largest economy, fell from almost 21.3 percent of GDP in 1992 (before Paul Martin as finance minister began his massive cutbacks) all the way down to 17.3 percent of GDP in 2003. This 4 percent drop amounted to about $57.85-billion in 2006. If Canada's social spending rose to match just the average spent by the 15 countries of the European Union, our increase in spending on education, health care, child care, the alleviation of poverty, labour training, public housing, etc., would be more than $95-billion.

While Canada's social spending fell, all the following countries *increased* their social spending: France to 28.7 percent of GDP; Germany to 27.6 percent; Austria to 26.1 percent; Portugal to 23.5 percent; Poland to 22.9 percent; Greece to 21.3 percent; the Czech Republic to 21.1 percent; Switzerland to 20.5 percent; the United Kingdom to 20.1 percent; Iceland to 18.7 percent; and Australia to 17.9 percent.

Slightly down from their previous levels of social spending, but still well ahead of Canada, were Sweden at 31.3 percent, Norway at 25.1 percent, Luxembourg at 22.2 percent, and the Netherlands at 20.7 percent. Bringing up the rear in social spending were the United States at 16.2 percent of GDP, Ireland at 15.9 percent, Turkey at 13.2 percent, Mexico at 6.8 percent, and 5.7 percent for Korea.

During the period 1990 to 2002, only six OECD countries decreased their social spending as a percentage of GDP, and Canada was one of them. And in a 2006 updated OECD report on social spending, most countries with a lower per-capita GDP than Canada were devoting more to social spending as a percentage of the economy than we were.

So, good old compassionate Canada is near the bottom of the list in social spending. What a bunch of hypocrites we've had in Ottawa. All of the above comparisons must be considered in relation to the steady stream of public opinion polls that clearly show most Canadians put social programs near the top of their list of priorities, far ahead of tax cuts, debt repayment, defence spending, and the economy. Just before the 2007 Conservative budget, a Strategic Counsel poll showed 50 percent of Canadians believed that increased government spending on social programs should be the most important priority, while only 19 percent supported tax cuts. Yet what we got were tax cuts, and later in the year even more tax cuts, with paltry benefits going to individual taxpayers.

It's worth noting that in the three years after Paul Martin became minister of finance, federal transfers to the provinces fell by $8.2-billion and federal social spending dropped by another $4.2-billion. By

2000/2001, federal program spending as a share of GDP had fallen to 11.6 percent, the lowest level in 50 years![2] Back in 1989, social spending accounted for 59 percent of total federal government spending. By 2007, it was down to only 49 percent.

Of course, as Ottawa chopped transfers to the provinces, the provinces, led by Ontario's Mike Harris and B.C.'s Gordon Campbell, slashed their own social spending. One result has been the pitiful levels of welfare payments we have had in this country for many years, as we shall see shortly.

The race-to-the-bottom impact of both the Canada-U.S. free-trade agreement (FTA) and North American free-trade agreement (NAFTA) is an important factor in all of this. And for the future, provisions in these "trade" agreements and Jean Chrétien's and Paul Martin's foolish promises to Quebec will combine to make important and desirable new national social programs such as a national pharmaceutical plan almost impossible.

As we've seen, in continental Europe, poverty has been reduced by 40 percent through enlightened social spending. In the Nordic countries and the Netherlands, the impact has been even greater. As for Canada, the United Nations Committee on Economic and Cultural Rights called it right when it released its third highly critical report about Canadian social policies in 2006. It made clear that "governments in Canada have not really committed to the recognition of social and economic rights as fundamental human rights." The UN committee complained about the fact that it was obliged to raise exactly the same points it highlighted in its 1993 and 1998 reports, and lamented the Canadian failure to implement its earlier recommendations. The Canadian NGO human-rights advocates who made submissions to the UN committee in Geneva argued, "In light of Canada's unrivalled economic and fiscal health, it is clear that Canada has chosen to permit the poorest people in the country to live at a level of misery that undermines their human dignity and violates their fundamental human rights."[3]

How could this have happened? As many of us had warned, the level playing field required by the FTA and NAFTA inevitably brought us down much closer to the uncaring American model. For as long as I can remember, Canadians have always taken for granted that social policy in our country has been, and will continue to be, very different from that of the United States. But this said, it is undeniable that thanks to the FTA and NAFTA, Canadian social policy today has moved closer to American standards and away from the more compassionate policies of most European OECD countries.

Next time you hear some Neanderthal describing Canada as a socialist welfare state, refer them to the pages you've just read. Next time you hear someone say that we don't have enough money for health care or education or social spending to help lift families out of poverty, tell them they should spend some time studying what almost all other developed countries do in terms of their social responsibility.

Let's compare what has been the situation in Canada for far too many years with what has recently happened in France, where a 2007 law makes housing an enforceable right, like health care and education. Beginning in 2008, the law will apply to the homeless, to single mothers, and to poor workers. By 2011, it will include all those living in poor-quality or unhealthy homes. And in Scotland, there is now a legally enforceable right to housing, committing government to supply housing for those who require it by 2012.

The excellent Canadian Centre for Policy Alternatives (CCPA) paper on tax revenues points out that

> Every just society must protect the vulnerable: children, the elderly and those with disabilities.
>
> In the Nordic countries, pensions replace 66.6% of the salaries of pensioners. In Canada it's 57.1%. In the United States it's 51%.[4]

Compare this with Finland, where it's 78.8%. All in all, for Canada another very shabby performance.

Widely respected political scientist Dr. Janine Brodie, in a paper delivered in Windsor in May 2007, sums it up: "One Canadian government after another has abandoned the vision of social citizenship, social security and social justice."

The Ottawa-based Caledon Institute of Social Policy commented on the Harper government's February 2007 budget: "The Budget could well have been named 'Opportunities Lost.' With a $19 billion price tag, never has so much been spent with so little result."

Returning to the subject of child care and early-learning programs, scores of studies from around the world have shown that countries with quality, affordable, universal early-learning programs perform better in a wide variety of ways than countries without such programs. The result is better students, fewer stressed-out teachers in elementary and secondary schools, and, overall, a more productive, innovative, and competitive nation. And study after study has shown that by far the best and most equitable child-care systems are properly financed universal public systems.

What is badly needed in Canada is one well-designed national child-care program, and not 13 different and mostly inadequate provincial and territorial schemes. Most European countries have long had national programs operating with great success, with big benefits to the social, educational, and behavioural performance of their children.

Despite all of this, the substantive 1984 promises made by the Mulroney government of a national early-learning and child-care program came to nought. Then, in 2006, the Harper government cancelled a national child-care agreement between Ottawa and the provinces that was the result of years of government negotiations and decades of advocacy by many informed and concerned groups. So today in Canada, more than 70 percent of mothers with pre-school children work, but fewer than one in five children under the age of six who have working parents have access to regulated child-care spaces. This compares to 60 percent in the United Kingdom and 78 percent in Denmark.

In our overall social spending, how utterly disgraceful it is for Canada, with the ninth highest GDP per capita, to be so uncaring and so uninterested in the welfare of its own men, women, and children compared to so many other developed countries.

That Stephen Harper has described Canada as "a Northern European welfare state in the worst sense of the term" tells you a great deal about what kind of person he really is.

Derek Burney, Brian Mulroney's chief of staff in the years leading up to the Free Trade Agreement, in an article published in the magazine *Policy Options*, and in the *National Post* on October 6, 2007, told Canadians that "None of the dire predictions about vanishing social programs ever materialized. . ."

What utter nonsense.

EMPLOYMENT AND
UNEMPLOYMENT IN CANADA

Employment in manufacturing and forestry in Canada is well down (forestry went from 89,000 jobs in 1995 to only 56,000 in the spring of 2007) and agriculture has had big job losses, falling from 512,000 in 1985 to 330,000 in 2007. But employment in finance, insurance, and real estate increased during the same years, from 694,000 to 1,059,000, and professional and related employment exploded from 425,000 to 1,118,000, while health and other public sector employment increased from 1.063 million to 1.839 million.

There has been huge growth in professional services and business services employment. This sector of our economy has gained more than 2.242 million jobs since 1992. Services now account for a solid majority of Canadian employment, and an increasing share of GDP.

This said, it's interesting to note that since early 2003 the growth of GDP has substantially outpaced the growth of employment. In 2006, there were 13.510 million Canadians employed full-time and 2.975 million employed part-time. Of all those employed, 8.727 million were men and 7.757 million were women.

While Canada has had a boom in well-paid jobs, there has also been a big growth in low-skill, low-paid jobs. In most Western European countries, low-paid jobs are somewhere between 8 and 12 percent, but in Canada they make up a big 21 percent of all jobs. Unfortunately, the large number of lost manufacturing jobs in Canada have been among the highest-paying in the country, year in, year out. About one in five jobs now pay less than $12 an hour. In addition, in many European countries, good public child care makes it much easier to go out into the workforce, and if necessary work at a relatively lower-paid job. In Canada, it's much more difficult, if not impossible, particularly in larger, more expensive cities.

Employment and unemployment rates, while very valuable indicators of the health of an economy, are in themselves incomplete if a number of other labour force factors are not also considered. For example, in some economies, part-time employment (which usually includes few benefits and much job insecurity) is a reflection of the underperformance of the economy.

In the spring of 2007, Canada's part-time employment was about 18.2 percent of total employment. In comparison, in 2006, the average for the G7 group of leading industrialized nations was 16.4 percent, and the OECD total 16.3 percent. The Czech and Slovak Republics have part-time employment percentages of only around 3 percent, followed by Hungary at 3.2 percent, Turkey at 5.8 percent, and Greece at 6 percent.[1] The U.S. rate of part-time employment in 2005 was only 12.8 percent. Statistics Canada reports that most of Canada's part-time workers were people who could not find full-time jobs, or those for whom child care presented a big problem.

Another useful way of looking at a country's economy is to measure self-employment as a percentage of total employment. Often, self-employment increases as the economy deteriorates and full-time jobs become scarce. Some contend, on the other hand, that a high rate of self-employment is often a sign of entrepreneurial skills.

In 2005, 21 OECD countries had a rate of self-employment higher than Canada's 15.5 percent. Greece was the highest at 30.2 percent, followed by Turkey at 28.7 percent, Mexico at 29.6 percent, and Italy at 25.5 percent. Luxembourg had the lowest rate, at 6.7 percent, Norway was at only 7.1 percent, and the United States was at 7.4 percent. Canada's self-employed rate was well above the G7 average of 10.2 percent, and also above the EU average of 14.5 percent and the OECD total of 14.4 percent.[2]

While job creation in Canada was good in 2007, a large number of these jobs have been in self-employment, most in the food services industry. For David Wolf, chief economist at Merrill Lynch Canada, "That's an awful lot of lemonade and hot dog stands."[3]

It's also important to note that despite their huge profit increases, private sector employment growth has been abysmal. In 2007, Canada's private sector employment increase was only a very poor 0.4 percent.

Let's look briefly at the relationship of free trade to employment. In the five years before the FTA came into effect in 1989, employment in Canada grew at an average annual rate of 2.9 percent. In the five years from 2001 to 2005, it grew at an annual average rate of only 1.84 percent.[4]

In 1989, the labour-force participation rate in Canada was 67.3 percent. In 2006, it was 63.6 percent. In the United States, it was lower at 63.1 percent. In 2006, in a list of 40 countries, Canada had the sixth highest labour force participation rate, higher than any other G8 country.

What is most amusing is the number of recent "think tank" reports attributing Canada's comparatively low level of productivity to the fact that Canadians don't like to – or aren't forced to – work as many hours as Americans. This is true if you look at the number of hours worked (see also the chapter on wages), but for the years 2004, 2005, and 2006, Canada's employment rate surpassed the U.S. rate. In our employment rate, in a list of 80 countries, we're in seventh place, tied with Denmark but ahead of the United Kingdom, Japan, and Germany, for example – in fact ahead of all G8 countries.[5]

When the FTA came into effect in 1989, unemployment in Canada was about 7.5 percent. It immediately began to climb, peaking at just under 12 percent in 1992/1993, before beginning a long fall back down to the pre-FTA levels.

Much has been said about Canada's recent low unemployment rates. But the 1990s saw the highest rate of unemployment in Canada of any decade since the Great Depression, and since the FTA came into effect, in the period from 1989 to 2006, Canada's unemployment rate was higher than that of the 30-nation OECD average every single year.[6] In the period from 1995 to 2005, 17 OECD countries had a lower rate of unemployment, while only nine were higher.

A word about the low U.S. unemployment rate. The very large American prison population in the United States and the increasingly large number of "discouraged workers" who have stopped looking for work serve to bring the rate down, while, remarkably, those who are classified as long-term unemployed are removed from the unemployment counts after six months!

It's very important to look at the prison population when considering American unemployment rates. The U.S. prisoner rate per 100,000 population was 725 in 2004, compared to the OECD average of 132.4 and Canada's rate of only 107. While Canada does well in this respect, all the following countries have rates lower than ours, beginning with the lowest and proceeding in ascending order: Iceland, at just 39, then Japan, Norway, Finland, Denmark, Sweden, Switzerland, Greece, Ireland, Belgium, France, Germany, Austria, Italy, and Turkey at 100, just below Canada's rate. In 2006, there were about 2.3 million men and women in U.S. prisons.

A few more words about the service sector in Canada, which now accounts for more than 70 percent of our economy and over 75 percent of all jobs. This sector has been responsible for most of Canada's employment growth in the past 10 years. It includes both high-paid jobs in financial services, transportation, information services, and telecommunications, but also lots of low-paid fast-food, janitorial, and other jobs.

What constitutes the service sector of our economy? Statistics Canada includes transportation and warehousing, finance and insurance, real estate, renting and leasing, the management of companies and enterprises, professional, scientific, and technical services, information and cultural industries, arts, entertainment, and recreation, administration, waste management, remediation services, educational services, health care, social assistance, accommodation and food services, public administration, and policing. In a list of the top 40 countries, Canada is in seventh place in our total services output,[7] but we invariably have a large trade deficit in services, reaching a record deficit of over $16-billion in 2006.

At this writing, Canada's unemployment rate is 5.8 percent, slightly higher than the OECD average of 5.7 percent. As Daniel Gross put it in the *New York Times,* "You can't use a low unemployment rate to pay a mortgage."[8] If, as we shall see shortly, distribution of income is badly skewed, average unemployment and income figures can present a very misleading picture.

6

OTTAWA'S UI/EI CASH COW

"Employment insurance is a myth."

A word about unemployment insurance, or, as it is now ridiculously called, "employment insurance." Before Brian Mulroney, Jean Chrétien, and Paul Martin went to work, more than 80 percent of unemployed workers in Canada received unemployment insurance; in fact, in 1980 it was as high as 86 percent. Today, it's down to only 40 percent. In Toronto, it's only 22 percent. In Ottawa, it's less than 21 percent.

Hundreds of thousands of unemployed Canadian workers now can't get EI and are forced to live on totally inadequate welfare with incomes far below the poverty line, thanks to the likes of the three above-mentioned prime ministers and premiers Mike Harris, Ralph Klein, and Gordon Campbell. In real terms, many Canadians on welfare are now receiving 45 percent less than equivalent rates of 10 years ago.

From 1962/1963 to 1982/1983, the unemployment insurance premiums received by Ottawa ran from a low of 3.2 percent of federal budgetary revenues to a high of 7.3 percent. But for the next 20 years they ranged from 9.3 percent of the federal government's revenues all the way up to 15.6 percent, a huge difference. For 16 of these last 20 years they exceeded 10 percent. From 1994 to the end of 2006, Ottawa had a massive employment insurance surplus of $51-billion.

Changes to the rules by Paul Martin when he was finance minister were destructive to the intent of the program. By 2006, only 53 percent of the unemployed even potentially qualified for benefits. But while payouts were being chopped, there were no equivalent cuts to premiums. While only 40 percent received benefits, 68 percent of the unemployed had been EI contributors. The result was that a social program intended to help the needy ended up being a very regressive tax used to pay down Ottawa's debt. Jean Chrétien and Paul Martin were widely praised for using the EI surpluses to cut the federal deficit.

Toronto Star columnist Thomas Walkom puts it all in the proper perspective:

> Unemployment insurance should be available to the unemployed. A minimum wage should bear some relationship to the cost of staying alive. Programs designed to reduce poverty should help the poor.
>
> These days, more people are working part-time at multiple jobs. Yet Canada's unemployment insurance system (which the federal government calls "employment insurance" to make it sound more positive) is available only to those who work in good, steady, full-time jobs – that is, to people who are almost never out of work.[1]

By May 2006, a task force designed to bring in recommendations to improve Canada's income security policies was direct and to the point: Ottawa had to reform unemployment insurance quickly to make it easier for out-of-work people to claim benefits. For David Pecaut, task force co-chair, "employment insurance is a myth."

In an editorial on February 25, 2007, the *Toronto Star* said this:

> Despite their sharp attacks on the fund while they were in opposition, the Conservatives have done absolutely nothing to reform the system.
>
> The benefit program must return to being a true insurance policy for those who lose their jobs, not a cash grab by the government at the expense of the most vulnerable in our midst.

How does Canada compare with other countries in terms of unemployment benefits? In a list of 28 OECD countries, we're way down in 22nd place when benefits are measured in terms of the replacement rates of previous earnings. Canada's rate is less than half that of Denmark, Finland, Israel, Germany, New Zealand, Austria, the United Kingdom, Belgium, France, Japan, and Australia.[2]

7

WELFARE IN CANADA

"An utter disaster"

"In reality, we do not care."

John Murphy is the former chair of the National Council of Welfare. He describes the present situation in Canada as "shameful and morally unsustainable." An editorial on welfare in the *Toronto Star* is to the point:

> Canadians pride themselves on being a caring and compassionate society. But, when it comes to the least fortunate in our midst – children included – we are anything but. We fail to provide them with the tools they need to help themselves, in the form of skills, training and access to childcare. Then we expect them to get along on incomes that don't reach half the poverty line.
>
> This country is rich enough to create a coherent national program that provides a decent level of income support to every family in true need. It is past time we recognized that, and acted on it.[1]

In their publication *Welfare Incomes* 2004, the National Council of Welfare wrote,

> Canadian welfare policy over the past 15 years has been an utter disaster.

Welfare incomes were further below the poverty line in most provinces in 2004 than they were in the late 1980s or early 1990s. Losses of 25 percent or more were reported in seven provinces.

And the council's *Welfare Incomes* 2005 said that

welfare incomes continued to decline in 2005, making life more difficult for the 1.7 million people – five percent of the population – forced to rely on welfare. Nearly half a million of those on welfare were children.

New Brunswick and Alberta had the lowest welfare incomes in 2005 for the four household types we looked at in each province and territory.

In 2005, welfare incomes were at the lowest point since 1986 in 20 scenarios.

When the peak year welfare incomes were compared to 2005 welfare incomes, some of the losses were staggering. In Alberta, the income of a single person decreased by almost 50 percent. In Ontario, a lone parent's income decreased by almost $6,600 and a couple with two children lost just over $8,700.

The income of a single person in Alberta with a disability was only 38 percent of the poverty line. And for a lone parent with one child, only 48 percent of the poverty line.

Across Canada, the council said, "No welfare incomes were remotely close to the poverty line, average incomes or median incomes."

As of early 2005, only Newfoundland and Labrador, Nova Scotia, New Brunswick, Quebec, and Manitoba had not clawed back any of the National Child Benefit Supplement .

In their 2005 report, the National Council of Welfare said,

Welfare incomes were woefully inadequate in 2005, as they have been every year since 1986, when the National Council of Welfare started tracking them. Welfare recipients are among the poorest of the poor and have to subsist on incomes far below what most people consider reasonable. They are so poor that they cannot access the resources that many of us take for granted – resources such as adequate housing, employment and recreational opportunities.

Welfare incomes were never high, but the recent declines demonstrate that governments are not interested in providing help to people who need it the most.

People don't turn to welfare because they want to; they turn to it because they have no other options. Who would choose to live on such a meagre income?

In 2005, for couples with two children, the gap between welfare incomes and the poverty line varied from an immense $19,553 in B.C. to $11,434 in P.E.I.

Commenting on the 2006 National Council *Welfare Incomes* report, the *Toronto Star*'s Thomas Walkom writes that the plight of those on welfare in Canada

> reflects the deliberate decisions of elected governments – presumably supported by the Canadian public at large – to purge roughly 1.7 million people consigned to welfare from our collective consciousness.
>
> It is shameful. It is pretty much criminal. Successive governments have gutted or eliminated much of Canada's vaunted social safety net.
>
> And increasingly employers prefer part-time or contract workers who can be fired at will and who are owed neither benefits or pensions.[2]

In March of 1995, there were some 3,070,900 men, women, and children on welfare in Canada. Since then, there has been a steady decline, down to 1,678,800 in March 2005. Pretty impressive, but certainly not all good news. While some of the decline can be traced to low unemployment rates, some of it stems from much more strict eligibility requirements. And those still on welfare are being punished by deliberate government policies in an unprecedented manner.

Economist Lars Osberg of Dalhousie University writes:

> Canadian society has become increasingly unequal in recent years, as incomes at the top have grown dramatically, while the least fortunate members of Canadian society have faced a substantially nastier economic reality. Although Canada's GDP per capita grew by 36 percent from 1986 to 2004, social assistance recipients in all Canadian provinces now have, after inflation, lower real incomes than comparable individuals did 20 years ago.
>
> Canadian society clearly does not care what happens to some of its citizens. Canada may have signed a series of international treaties on human rights that declare adequate housing to be a basic human right – but in reality we do not care.
>
> Canada now has both more "monster homes" and more homeless, while in recent years there has been a substantial erosion in social safety nets.[3]

In 1995, Paul Martin and the federal government reduced transfers to the provinces and introduced a lump-sum transfer to include social assistance, health, and post-secondary education. Where once there had been shared-cost accountable programs, there were now different levels of programs across the country, with the provincial governments having full responsibility.

So what did the provinces do? They promptly brought in tough rule changes that made benefits much more difficult to obtain, while at the same time they cut the real value of social assistance benefits. By 2000, recipients were typically getting only about 30 percent of the Statistics Canada low-income cut-off levels.

In Ontario, thanks in large degree to Mike Harris's drastic cutbacks, the number on welfare dropped from just over 500,000 in 1994/1995 to some 200,000 in 2005, and the provincial government clawed back up to $1,463 from low-income families receiving the federal government's child benefit supplement. Meanwhile, welfare cuts and inflation cut the purchasing power of benefits by roughly 40 percent.[4] Individuals who have to depend on social assistance receive only $536 a month, almost 70 percent below the Statistics Canada low-income cut-off. And while Ottawa sends a monthly cheque for $162 per child to the lowest-income families, Ontario claws back over $121 from children whose parents receive the meagre social assistance.[5] In March 2007, Premier Dalton McGuinty introduced a new Ontario child benefit which, by 2011, should end that province's clawbacks.

Let's turn to B.C. In the words of the respected Vanier Institute of the Family, "The situation in British Columbia is startling," with a sharp drop in welfare recipients and a huge increase in child poverty. In B.C., the province with the highest child poverty rate, 90 percent of those who applied for welfare in 2001 were successful. After the Campbell government's brutal tightening of the eligibility rules, by 2004 only 51 percent were granted assistance. Huge numbers of men, women, and children with a genuine need were left out of the system. Between 2000 and 2005, in a booming economy, child poverty in B.C. increased by almost 8 percent.

As for wealthy Alberta, one can only have bitter contempt for the egregiously low levels of social assistance in a province with no deficit or debt, enormous surpluses, low taxes, and punishingly inadequate welfare programs.

Then there's our ridiculous federal and provincial tax systems. In a May 2006 editorial, the *Globe and Mail* said,

> As it stands, the nexus between welfare and work is a mess. Provincial and federal systems interfere with each other, often to disastrous effect. Suppose a single parent with one child somehow scrounges an extra $10,000. The marginal tax

rate for those extra dollars of income is an incredible 78 percent. And that does not include the potential loss of benefits such as subsidized prescription drugs and housing.[6]

Yes, difficult as it may be to comprehend, that was 78 *percent!*

As TD Bank Financial Group economists Don Drummond and Gillian Manning pointed out in a 2005 paper, the system in Ontario insanely penalized anyone on welfare who begins to earn an income: "Under the original Harris scheme, a welfare recipient who earned a dollar could lose more than a dollar in benefits." And even under changes made by the McGuinty Liberal government in Ontario, a social assistance recipient earning a dollar loses 50 cents in benefits, the equivalent of a 50 percent marginal tax rate. In the 2003 provincial election campaign, the Liberals promised to end the Mike Harris clawback of the $122-per-month-per-child national child benefit supplement, a supplement that has frequently meant the difference between food and no food. Four years later, at the time of this writing in 2007, they had still failed to do so.

Tom Walkom writes:

> Welfare systems are being asked to fill the gaps left by the lack of affordable child care, dental care and drug coverage. They are also being asked to fill in for an employment insurance system that for reasons both deliberate and circumstantial no longer covers most people out of work.
>
> Economists Drummond and Manning suggest two new federal programs: an earned income supplement for the working poor (in effect a wage subsidy for employers) and a refundable tax credit for the very poor.
>
> What this means is that poor people would file tax returns even if they have no earned income, and the government would send them cheques. As such, it is a variation on the old guaranteed annual income scheme, an idea that at different times has had currency with both the left and the right.[7]

As Don Drummond has shown, after Mike Harris went to work on Ontario's welfare system, a single welfare mother with one child could lose as much as 92 percent of the money she would make if she worked 64 hours a week at $7.45 an hour. Because of lost benefits, she would net only $40 extra a week, barely enough to cover the cost of taking public transit to work. Even with some modest reforms by the provincial Liberal government, the same single mother would still face a marginal tax rate of almost 60 percent and, as the *Toronto Star* has pointed out, if she worked even longer hours to try to escape welfare, her tax rate would increase to 71 percent.

If this isn't ridiculous, I don't know what is.

The gulf between Canada's big business leaders and most Canadians on the subject of social policy is enormous, as it is in so many areas covered in this book. Opinion polls show that the vast majority of Canadians continues to support our long-standing social policies and has little or no interest in moving closer to American standards and values. On the other hand, our corporate leaders see their interests best served through even more tax cuts, and they envy their American peers who obviously have little or no interest in effective social policies of the kind found in almost all the other Western democracies. Despite public opinion, successive Canadian governments have chopped federal program spending to levels few ever imagined or supported. (See the chapter on government in Canada for details.)

How has big business reacted? One poll of Canadian CEOs asked, "Where should provincial governments reduce spending?" In response, 43 percent said social programs, more than three times any other category named.

Imagine what we could do if our social spending was raised to only the average level of the other developed Western democracies. Think of what we could do in health care, post-secondary education, cutting back child poverty, public housing, school lunch programs, not to mention research and development for renewable energy.

Where would we get the money? Or, to put it another way, how can we become a more normal developed nation? Some of the answers to this will be found in what you've already read, and more in the tax section in this book. Before you get to that, let me steal from the Canadian Centre for Policy Alternatives's excellent publication *The Social and Economic Costs of Taxation:* "A very famous U.S. jurist, Justice Oliver Wendell Holmes, once remarked, 'Taxes are what we pay for a civilized society.'"

Despite so much right-wing propaganda to the contrary, the fact is that total tax revenue in Canada as a percentage of GDP put us far down in 21st place among the 30 OECD nations, well below the OECD and EU averages. And this was before the Harper government's big 2007 tax cuts.

I am not for a moment suggesting that we immediately bring our social spending all the way up to the average of the EU countries. But I am suggesting, in no uncertain terms, that we should be thoroughly ashamed of our own pitiful level of public social support and that we have lots of room to fix it.

What could we realistically do to combat poverty in Canada? In brief, we could raise minimum wages at least to $10 an hour to help provide a more decent standard of living for our 650,000 working poor. We could provide an earned-income supplement which takes into consideration regional cost-of-living factors. We could stop taxing people with very low incomes. We could return to a much more realistic and effective unemployment insurance program. We could stop the provincial clawbacks of child benefits and increase the supplement for low-income parents. We could develop an affordable rental housing program. We could increase social assistance benefits and expand job training. We could develop a national early-education and child-care program and provide more funds to help pay for university and college tuitions and expenses. We could substantially increase such public expenditures, which in 2004 were well below the standards of the Nordic countries, the Netherlands, Germany, France, Switzerland, Spain, and Italy, both in quality and as a percentage of GDP.

As for the Harper government's 2006 plans for more assistance for disabled children, the media in Canada have somehow been blind to the inadequacies of the plan. Hillel Goelman, of the University of British

Columbia, spelled it out nicely, however, in a letter to the *Globe and Mail*: "Tax breaks do not provide more hours of physiotherapy, tax breaks do not provide more trained, early-intervention therapists, tax breaks help only wealthy parents who have enough income that they can benefit from deductions."[8]

Would ending the clawbacks of the national child benefit supplement (NCBS) make a difference to families on social assistance? Just under half of families who go to Toronto's Daily Bread Food Bank say that the extra $122 a month per child would mean they would no longer have to depend on the food bank. By 2005, some 174,250 families with almost 281,000 children continued to have their welfare or child benefits clawed back by the Ontario government.

In the summer of 2006, a new Statistics Canada study showed that only 26 percent of some $3-billion paid out by the federal government in GST rebates went to low-income families, while some of the balance went to families earning up to $100,000 or more. Once again, if this isn't ridiculous, I don't know what is.

Today, all across Canada, rent eats up the vast majority of any welfare or other benefits the poor receive. The very small amount left for food, clothing, transportation, school supplies, hygienic supplies, utilities, and so on, is hopelessly inadequate.

We should be ashamed. Very, very ashamed.

IMMIGRATION AND EMIGRATION

I n 1981, Canada took in 127,000 immigrants. Ten years later, in 1991, it was 221,000, and in 2001 it was up to 253,000. In 2005, the number was up again, to over 262,000. The same year, some 36,000 emigrants left Canada, well below the 48,000 average figures for 1999 to 2001, leaving us with a large plus migration rate.

Of the 30 OECD countries, in the period from 2000 to 2005, only Italy, Ireland, and Spain had a higher plus migration ratio than Canada. Canada's ratio is better than such countries as the United States, Australia, Germany, New Zealand, Sweden, and Norway, and far ahead of such countries as Japan, Finland, France, the United Kingdom, and Belgium. Moreover, in recent years, Canada's plus migration ratio has been increasing, from 3.8 per 1,000 in 1991, to 6 per 1,000 in 2004.

How do immigrants to Canada compare with immigrants to other OECD countries when it comes to the percentage of all those with a post-secondary education? The answer is very well indeed. In Canada, such educated immigrants make up more than one in five of all persons with a post-secondary education. Only Luxembourg, at just over one third, and Australia, at just over one quarter, do better. Of the immigrants who arrived in Canada between 1996 and 2001, more than 52 percent had a university education, compared to the Canadian-born average of only 21 percent.

Canada's record in this respect is much better than all the other OECD countries, and much better than all the G7 countries, including the United States.[1]

"Brain drain?" Nineteen OECD countries have a higher percentage of persons with a post-secondary education leaving to emigrate to other OECD countries, and emigration of such persons from Canada is far below the OECD average.[2] As I have pointed out often elsewhere, the brain drain mythology promoted by the likes of Conrad Black and the *National Post* has been grossly overstated. There is abundant evidence that rather than a brain drain, Canada has a huge brain *gain,* every year, year after year. Between 1986 and 2006, emigration from Canada fell by 24 percent.

In 1930, there were 1,310,000 Canadian-born people living in the United States. By 1960, there were only 953,000. By 1970, only some 812,000. And by 2000, just 678,000. In 1990, Canada had a net gain of 1.5 million people with post-secondary education. Ten years later, the net gain was 2.25 million people.

In 1981, 16.1 percent of Canada's population was foreign-born. By 2004, it was 18.0 percent.[3] In 1981, only 5 percent of Canadians were "visible minorities" other than aboriginal persons. By 1996, this was up to 11 percent, and by 2006 it was 16 percent. Statistics Canada has forecast that by 2017 about one in five Canadians could be from a visible minority. South Asians and Chinese will continue to be the largest visible minority groups.

In February 2006, Dr. Ivan Fellegi, the head of Statistics Canada, reported that about 35 percent of the Canadian population are either first-generation immigrants or the children of immigrants. While Australia had almost 25 percent of their population of immigrant stock by 2000, Canada was next at almost 19 percent, followed by the United States at 12.4 percent, France at 10.6 percent, Germany at 9 percent, and the United Kingdom at 6.8 percent. In Italy it was only 2.8 percent.

Today, about 20 percent of Canada's population is foreign-born, compared to the OECD average of only about 8 percent. Only New Zealand, Switzerland, Australia, and Luxembourg have higher foreign-born percentages. Canada has the fifth highest percentage of foreign-born residents in the OECD. By the end of 2006, foreign-born people in Canada accounted for almost one in five of the total population, the highest proportion in 75 years.

In 2005, about 53 percent of the 262,236 permanent residents admitted to Canada came from the Asia and Pacific regions, just under 19 percent came from Africa and the Middle East, 16 percent from Europe, 9 percent from South and Central America, and 3.5 percent from the United States.[4]

In 2006, the number of U.S. citizens who moved to Canada, 10,942, was at a 30-year high. Jack Jedwab of the Association of Canadian Studies says that the Americans who are coming to Canada "have the highest level of education. They're coming because many of them don't like the politics [in the United States], the Iraq War and the security situation in the U.S. By comparison, Canada is a tension-free place. People feel safer."

The highest percentage of immigrants to Canada in 2005 came from China, 16 percent, followed by 13 percent from India, 7 percent from the Philippines, and 5 percent from Pakistan.

In 2007, there was a terrible and inexcusable backlog of some 800,000 applicants waiting to come to Canada. Some potential immigrants have been kept waiting up to five years before being given a decision on their application. And even if granted citizenship, many professionals must wait years to have their working credentials recognized.

About 80 percent of new immigrants settle in Canada's big cities. And, as has often been pointed out, because of our regrettable reluctance to accept foreign credentials, "in Toronto, Montreal or Vancouver the taxi driver is often better educated than the passenger."[5] In 2001, almost 44 percent of the people in Toronto were foreign-born, just under 38 percent in Vancouver, and just over 18 percent in Montreal. By comparison, Miami had 40 percent, London 25 percent, and New York City only just over 24 percent. Looking ahead, Statistics Canada projects that by 2017, 51 percent of the population of Toronto will be visible minorities, 49 percent of Vancouver's population, 28 percent of the Ontario parts of Ottawa-Gatineau, 24 percent of Calgary, and 23 percent of Windsor.

In the United States, things have changed. Migration from around the world to the United States peaked in 2000 and has declined a great deal since, down roughly 25 percent, from about 1.5 million in 1999 and 2000. The number of legal permanent residents entering the United Sates was only 647,000 in 2000, but that fell to just 455,000 in 2004. The same year, the number of "unauthorized" immigrants was estimated at 562,000.[6]

A major unfortunate change in Canada has been the large percentage of immigrants who now live in poverty. While employment rates for the Canadian-born have been increasing, from about 73 percent in 1981 to 81 percent in 2001, they have been dramatically falling for recent immigrants, from about 77 percent to just over 68 percent. By 2004, some 34.5 percent of immigrants to Canada who had arrived here within the previous 10 years were classified by Statistics Canada as "low income."

In January 2007, Statistics Canada reported that

> the economic situation of new immigrants to Canada showed no improvement after the turn of the millennium – despite the fact that they had much higher levels of education and many more were in the skilled immigrant class than a decade earlier.
>
> In 2002, low-income rates among immigrants during their first full year in Canada were 3.5 times higher than those of Canadian-born people. By 2004, they were 3.2 times higher.
>
> These rates were higher than at any time during the 1990s.
>
> Among new immigrants aged 15 and older, the proportion with university degrees rose from 17% in 1992 to 45% in 2004. And the share in the economic skilled immigrant class increased from 29% to 51%.
>
> One in five recent immigrants were living in low income at least four years during their first five years in Canada. This was more than twice the corresponding rate of around 8% among Canadian-born people.
>
> Among those who arrived in 2000, 52% of those in chronic

low income were skilled economic immigrants. About 41% had a university degree.[7]

According to the Royal Bank of Canada, if foreign-born workers were as successful as Canadian-born workers, personal income in Canada would be about $13-billion a year higher. But of course that's an irrelevant figure if skilled foreign workers can't find a decent job because their credentials aren't recognized, their language skills are lacking, or because of discrimination.

While immigrants aged 25 to 54 have been more likely to have at least a bachelor's degree (36 percent) compared to Canadian-born men and women (22 percent), nevertheless their unemployment rate in 2006 was four times the rate for Canadian-born university-educated workers, while for immigrants with a graduate degree it was over five times higher (12.4 percent compared to 2.4 percent).[8]

Canada's current fertility rate of 1.5 children per woman means Canadians are not replacing themselves. This means an aging of the population where within 20 years there will be more deaths than births. It also means that unless immigration increases, our population will decline. Moreover, the median age will increase from just under 39 years to almost 44 years by 2025, and our natural growth in our labour force will come nowhere near to meeting demand as soon as 2016. Statistics Canada reports that in 2006 the proportion of children was the lowest ever recorded.[9] Many other countries face similar problems. Some forecasts are that Italy, Spain, and Japan will lose over 20 percent of their populations over the next four decades and the population of the European Union will drop by a huge 100 million.

By 2005, new Canadians were already making up some 70 percent of the growth in the labour force. Some estimates say that by 2011 they will account for 100 percent of the growth in the labour force. Other estimates suggest that by mid-century Canada will need almost four times as many immigrants to maintain our labour force growth rate of the past few years.

The C.D. Howe Institute forecasts that even with an immigration increase to 320,000 a year, the ratio of elderly to those of working age will almost double by 2035. The Conference Board of Canada suggests that within 20 years, the ratio of workers to pensioners will move from five to one down to three to one. The impact on our health-care system and pension plans will be profound.

It's interesting to look at the public opinion polls relating to immigrants. Almost 80 percent of Canadians believe immigrants have a good influence on our country. In Australia, it's only 52 percent. In the United States and Britain, only 43 percent. Compared to most other Western countries, the anti-immigrant sentiment in Canada is much smaller. An August 2005 poll showed that fewer than one in three Canadians believed that we accepted too many immigrants, and most Canadians felt we were a better country because of multiculturalism.

When, in 2005, federal Immigration Minister Joe Volpe called for an increase in the number of immigrants to more than 340,000 annually by 2010/2011 a *Globe and Mail* editorial said, "Canada may be the only country in the world where it's seen as good politics before an election to call for a major *increase* in the number of foreigners who come in."[10]

It's interesting to note that on average 84 percent of eligible immigrants to Canada become Canadian citizens. Also interesting, and most disappointing, is the fact that only 32 percent of U.S.-born immigrants who have lived in Canada for more than 30 years have become citizens. Thirty years!

While in the 1970s, 30 percent of immigrants said they knew neither English nor French, by the 1990s this had grown to 45 percent. With such high numbers, it's obvious that lack of adequate language skills is a major factor in poor employment and income levels. Yet 80 percent of recent immigrants say they have a strong sense of belonging to Canada.

In 2006, Solutions Research Group reported that 86 percent of immigrants believed that Canada was "the best place in the world for me to live," and 92 percent of second-generation Canadians agreed. Between 72 percent and 86 percent agreed that Canada's health-care and education systems were better than in their country of origin, as were safety and security.

One of the most pleasurable things I have ever done is to have acted as a citizenship judge in ceremonies swearing in new Canadian citizens. The last time I did this, there were 97 men, women, and children from 30 different countries. In particular, I remember the smiling faces of an oncologist from Boston and his family and a computer programmer from Siberia and her family. Great happiness at becoming Canadian citizens filled the room. There were smiles everywhere, and many tears of joy as we all sang "O Canada."

I was very proud to have had this touching opportunity.

It's worth noting that Canada has the highest proportion of immigrants in parliament of any country.

PART THREE

WAGES IN CANADA

"Economic justice is off the radar screen."

W hy are so many children poor in an affluent nation like Canada? One major reason is that their low-paid parents have to spend so much of their income on rent and utilities, leaving little even for such basics as food and clothing, and there is no public housing available to them.

A Queen's University study by Richard Chaykowski found that millions of jobs in Canada don't pay enough to keep a family out of poverty. One in six Canadian workers earns less than $10 an hour, one in 25 earns the minimum wage or less. Aside from the large number of low-paid full-time jobs, there have been growing percentages of part-time casual and temporary jobs with irregular hours, poor working conditions, few benefits, and little chance for advancement.

Chaykowski says that a third of all working Canadians earned less than poverty wages. Carol Goar of the *Toronto Star* puts it this way: "As things now stand, millions of workers are not covered by employment standards and are not eligible for jobless benefits. And temp agencies can skim off half a client's wages. Workers can be rented and returned like spare parts."[1]

Let's take a closer look at how workers in Canada have been doing in our growing economy. In 1995, the industrial aggregate of hourly earnings was $15.05. Ten years later, it was $18.17, representing an average annual increase of just over 31 cents, or 2 percent, a year. During the same years, the average inflation rate was 2.24 percent. So, while the economy has been booming in recent years, Canadian workers haven't exactly struck it rich.[2] During the 20 years from 1987 to 2006 inclusive, the annual increase in the consumer price index was greater than the increases in wage settlements in 10 of those years, and in two of the remaining 10 years they were identical.

One interesting way of looking at wages is as a percentage of GDP. In 1992, before taxes they were 55.4 percent. In 1993, 55 percent. Since then, they've gone downhill, to only 50.2 percent in 2005. The 5.2 percent difference worked out to almost $7.3-billion in 2005. (In 2006, wages increased marginally to 50.7 percent of GDP, still far below the level of most previous years.)

In sharp contrast, it's revealing to look at corporate profits as a percentage of GDP during the same years. In 1992, before taxes they were 4.7 percent of GDP; in 1993, 5.7 percent. But in 2006, they were up to 13.9 percent of GDP, the highest in history.

Another way to measure wages is to compare contractual wage settlements to increases in GDP. For the period from 1996 to 2005, hourly settlements averaged just 2 percent annually. Meanwhile, average annual GDP increases at market price averaged 5.4 percent. Once again, while the economy has been doing well and profits have been soaring, workers are showing little or no improvement in their earnings.

In a different measure, the Vanier Institute of the Family reported in February 2006 that since 1990 the average after-inflation increase in hourly earnings for Canadians had been only a tiny 10 cents. And in a 2004 study of 11 countries, the consulting firm KPMG said that Canada had the lowest labour costs among all the countries in their survey.[3] Great for corporations, but hardly something for workers to rejoice about.

It's interesting to compare wage increases in Canada with those in other developed countries. In the 10 years from 1996 to 2005 inclusive, wage increases in Canada were below the average wage increase of the other 29 OECD countries in eight of the 10 years.

Statistics Canada defines "low wage jobs" as jobs paying less than $10 an hour (in constant 2001 dollars). In 2004, over 1.3 million full-time

Canadian workers had low-wage jobs. Women are twice as likely to be in low-wage jobs as men. Dennis Raphael of York University says that in contrast to Canada, only 5 percent of Finnish and Swedish workers earn low wages, and, as we've seen, their poverty rates are far lower than Canada's. Raphael says, "In essence, the single best predictor of the number of people living in poverty in a nation is the number of people earning low wages."[4] Raphael also says, "Among developed nations, Canada has the second highest percentage of low-paid workers . . . exceeded only by the U.S. Our minimum wages are among the lowest."[5]

Economists Arthur Donner and Douglas Peters, in a May 2006 *Toronto Star* op-ed piece, are to the point:

> To a far greater extent than in the past, the issue of economic justice and income distribution is off the radar screen of our politicians, leaving huge numbers living in poverty.
>
> We are struck by the way in which the minimum wage was deliberately allowed to fall despite the fact that the minimum wage plays a key social and anti-poverty role.[6]

From 2000 to 2005 inclusive, the share of employees working for minimum wage in Canada varied from a low of 4.1 percent to a high of 4.8 percent. In 2005, the highest provincial percentage was in Newfoundland and Labrador, at 6.8 percent, the lowest in Alberta at 1.5 percent.[7] That same year, some 587,000 men and women worked at or below the minimum wage in their province, which ranged from a low of $6.25 an hour in Newfoundland and Labrador to $8.00 an hour in British Columbia. By 2007, Newfoundland and Labrador as well as Alberta and New Brunswick had minimum wages of $7.00 an hour, while the other provinces were between $7.15 and $8.00, the rate for Ontario and B.C.

Women made up just over 60 percent of all minimum-wage workers. This works out to about one in 19 women workers, compared to one in 30 men.

Again, how do minimum wages in Canada compare with those in other developed countries? If we compare the statutory minimum wages plus social-security contributions for average workers in 17 OECD countries, 11 have higher combined wages and benefits than Canada. The only countries lower than Canada on the list are Poland, the United States, the Czech Republic, Japan, and South Korea. In July 2007, the United States increased the minimum wage to $5.85 an hour, the first increase in a decade.

For the most part, especially considering inflation, minimum wages in Canada have improved very little in recent years. In some provinces, they have gone backward in real purchasing power. After many years, the Ontario minimum wage finally went up from $6.85 an hour in 1995 to $8.00 in 2007. But in high-cost cities such as Toronto, eight dollars an hour is hardly enough to come close to a decent standard of living. Ontario is not an exception. In no Canadian province are minimum wages close to the poverty line.

In Ontario, almost 1.2 million people have to live somehow on wages of less than $10 an hour. And even with the 2007 increase of 25 cents, in real terms they were actually earning 10 percent less than a decade earlier. A 2007 Canadian Centre for Policy Alternatives (CCPA) report showed that if the real wages of workers in Canada had increased at the same pace as productivity, average annual salaries would be some $10,300 higher.

Now let's look at how Canadian workers have made out compared to their CEOs.

In the fall of 2006, University of British Columbia economist Robert Evans, writing in *Healthcare Policy*, pointed out that between 1976 and 1990 average per-capita income hardly changed, but the top 0.01 percent of earners saw their incomes more than double. Evans blames the increasing wealth and power of the very rich for the increasingly unfair elite agenda of lower taxes for high-income recipients and lower social spending. Meanwhile, CEOs are firing well-paid workers and contracting out their jobs at lower pay with fewer benefits.

In the above-mentioned report from the CCPA, economist Hugh Mackenzie showed that in 2005 the average top-paid CEO made just under $9.06-million and the top 100 ranged from $2.87-million to over

$74.82-million. Meanwhile, the average Canadian worker earned about $38,000 in a year, and the average person working for minimum wage earned $15,931 a year.

Mackenzie illustrated the difference in annual income by working out that the average top-paid CEO had equalled the average Canadian's yearly earnings by 9:46 a.m. on January 2. And as for the country's highest earner of all,

> Canada's highest-paid CEO in 2005 would barely have had time for morning coffee on New Year's Day before matching average earnings for Canadians for the whole year. His pay passed the Canadian average at 10:04 a.m. on January 1st.
>
> The highest-paid CEO makes as much as a small town – 1,969 people – working at the average of wages or salaries, or 4,686 people working full-year at the minimum wage.

One Conference Board of Canada survey in 2005 indicated that pay for corporate directors soared 41 percent over two years, far more than 10 times the increase in the average wage of workers during the same period. Of the 50 highest-paid executives in 2005 (salary, plus bonus, plus shares, plus options, etc.), all received more than $9-million in compensation. Seven received more than $20-million. Hank Swartout, CEO of Precision Drilling Trust, received over $74.8-million; Hunter Harrison, CEO of Canadian National Railway, was awarded $56.2-million; Frank Stronach of Magma $52.5-million; Mike Zafirovski of Nortel over $37.4-million; and John Hunkin, former CEO of the CIBC, received a total of just under $29.5-million.[8] Most of these represented enormous increases from their compensation of a year earlier. In 2007, James Balsillie and Michael Lazaridis received $92.5-million, André Desmarais and Paul Desmarais Jr. received $40.2-million, and Glenn Murphy had to get by with only $34.4-million.

We've already seen the tiny percentage wage increases Canadian workers have received. Let's look at some recent CEO percentage increases: Denis Turcotte of Algoma Steel, 183 percent; Marc Tellier of Yellow Pages, 163 percent; Nancy Southern of Canadian Utilities, 135 percent; Donald Lang of CCL Industries, 138 percent; Clayton Riddell of Paramount Resources, 101 percent.

According to the *Globe and Mail*, Canada's chief executives saw their average compensation increase by 39 percent in 2005. Adam Zimmerman, the retired CEO of Noranda Forest, has been one of only a small handful of Canadian executives to speak out against the excessive levels of corporate compensation. In an October 2005 speech in Toronto, he said, "These huge remuneration schemes are the root cause of all that is going wrong in the world."

The major banks are the "princes" of the compensation world, and other companies follow them. The CEOs of Canada's six biggest banks were paid a total of $53-million in 2006. Zimmerman said that CEO pay should be put in perspective with the salaries of normal employees. In one CCPA study, the 100 highest-paid chief executives made 218 times as much as the averge full-time Canadian worker. "Were it not that they do it within the law," Zimmerman said, "I would think executives are stealing money."[9]

Hand in hand with growing income and wealth inequality have come increasing corporate concentration, media concentration, and globalization (where transnational corporations oppose social reforms which might increase their costs). Dennis Raphael and Toba Bryant write:

> Since the mid-1970s a fundamental change in the operation of national and global economics has occurred. The increasing ability of transnational corporations to easily shift investments across the globe pressures national governments to accede to demands to reverse reforms associated with the welfare state. International trade agreements weaken national identities and nationally based labour unions. Trade is now international, but unions are nationally based. With such a power shift, business has less need for political compromises

with labour and even governments. Worker power and the ability to negotiate better wages and benefits are weakened.[10]

Now that we've looked at wages, let's turn our attention to the remarkable figures that demonstrate the distribution of income and wealth in Canada.

THE DISTRIBUTION OF INCOME IN CANADA

"A substantially nastier economic reality"

"While the poor have become poorer,
the rich have become much richer."

W e've already seen how poorly the average worker in Canada
has done in recent years. Now let's take a broader look at how
incomes, including family incomes, break down; that is,
who's gaining and who's been missing out.

From 1990 to 2000, families in the highest 10 percent of income
earners in Canada increased their incomes by an average of 14.3 percent.
Middle-income families, however, saw their incomes increase by only a
tiny 0.3 percent, and those below middle-income levels *lost* on average
0.7 percent of their incomes. Over the entire 16 years from 1984 to 2000,
the average after-tax income of the poorest fifth of Canadian families
increased by less than $1,000!

Looking at the incomes of all Canadians, not just families, in 1990
the average personal income of the top 10 percent was $161,460. By
2000, it was $185,070, a gain of $23,610. There was quite a different story
for the lowest 10 percent, whose average earning in 1990 was only
$10,341. Ten years later, that had increased by only $80 a year. In other
words, the top 10 percent increased their earnings by over 295 times the
lowest 10 percent!

In October 2000, Paul Martin, who was then finance minister, said,
"We must work towards reducing the gap between rich and poor. . . . We
have always said we must work towards reducing this gap." Let's see the
results of what Paul Martin and the Liberals "have always said." Here's

what happened after Mr. Martin and Mr. Chrétien went to work. Four years after Martin's promise to reduce the gap, in 2004, the highest 20 percent of Canadian families earned 42 percent of all market income, while the lowest 20 percent earned only 3.6 percent.

Back in 1989, the year of our now infamous House of Commons resolution on child poverty, and the first year of the FTA, the bottom 40 percent had over 17.2 percent of income. Measured by quintiles, in 2004 the top 20 percent of families had 46 percent of all market income while the bottom 20 percent had a tiny 3.6 percent.[1]

In a July 2004 paper, economists Emmanuel Saez, of the University of California at Berkeley, and Michael Veall, of McMaster University, make some important points:

> Over the last 20 years, top income shares in Canada have increased dramatically, almost as much as in the United States. This change has remained largely unnoticed because it is concentrated within the top percentile of the Canadian income distribution and thus can only be detected with tax return data covering very high incomes.
>
> The upturn during the last two decades is concentrated in the top percentile (whose share increased from about 7.5 percent in the late 1970s to 13.5 percent in 2000 . . .).[2]

One reason was an "unprecedented surge in the pay of top corporate employees." Neil Brooks, noted York University tax specialist, says that the top 1 percent of Canadians doubled their share of the total national income between 1980 and 2000.

Taking a longer look, from 1980 to 2005, Statistics Canada has calculated that during this period the richest 20 percent increased their incomes by 24 percent. During the same years, the average income of the poorest 20 percent increased by only 4.9 percent, or $600; a grand total of a pathetic $24 a year.[3]

Statistics Canada put it this way in the spring 2005 issue of *Perspectives on Labour and Income:*

> During the 1990s, gains associated with economic expansion
> in Canada went mainly to higher-income families.
>
> While incomes among the richest 20 percent of families
> were rising by about 10 percent, they stagnated among the
> poorest 20 percent between 1990 and 2000.
>
> In addition, the mid-1990s saw an unexpected increase in
> the low-income rate. As unemployment fell, the low-income
> rate continued to rise.
>
> This may be attributed to earnings difficulties among
> poorer families and declining social transfers.[4]

"May be"? If you've already read the chapters in this book on poverty, wages, and social programs you'll know that there's no "may be" about it.

In the words of University of British Columbia economist and distribution-of-income specialist David Green, "There has been a notable rise in market income inequality because of declining real incomes at the bottom and rapidly rising real incomes at the top."[5] Green quotes other Canadian economists who have shown that real market income fell for the lowest 25 percent of families during the 1980s and 1990s while "the share of total income going to the top 1 percent of individual earners rose sharply over the 1980s and, then, even more dramatically in the 1990s."

When the final figures for 2000 to 2007 are in, the results are certain to be far more dramatic, and, for almost all Canadians, far more shocking. Except, of course, for the top 1 percent. Statistics Canada has already confirmed that income inequality continued to grow during the first half of 2007, with the income difference between the top and bottom fifths of families up by another third.

Mind you, the figures you have been reading do not include capital gains, so the actual imbalance is even greater. Much, much greater.

As bad as all of this seems, it's actually even worse. Economists Marc Frenette, David Green, and Kevin Milligan, in a Statistics Canada paper,[6] make a number of important points. They find that after-tax inequality levels in Canada are substantially higher than previously thought, and that previous sources under-represented both very low and very high incomes, producing an "underestimation of the level of inequality" in Canada, and "a substantial long-term increase in disposable income inequality."

Their study puts average total income in the lowest decile in the year 2000 at only some $6,000, compared to previous estimates of almost $8,000. At the same time, the average total income in the top decile was about $98,000, about $7,000 more than previous estimates.

Economist Lars Osberg of Dalhousie University, in a *Policy Options* article,[7] writes, "While the poor have become poorer, the rich have become much richer. The real income (as reported for income tax purposes) of the top 1 percent of Canadian income earners increased by 65.7 percent from 1986 to 2000."

How do we Canadians do in terms of income distribution compared to other countries? In the 2007/2008 edition of the United Nations's *Human Development Report*, when you compare the incomes of the richest 10 percent to the incomes of the poorest 10 percent, 38 countries had a more equitable distribution of income. While Canada was 39th on the list, the U.S. was way down in 71st place.

If you compare the incomes of the top 20 percent with the incomes of the poorest 20 percent, Canada is in 34th place. Some of the countries that do much better than Canada are Norway, Sweden, Japan, Finland, Austria, Denmark, Germany, the Czech Republic, Hungary, Slovakia, and Croatia. In the *OECD Factbook,* 2006, some other countries with better-than-average equitable distributions of household disposable income include the Netherlands, Luxembourg, and Switzerland. The worst records are for Mexico, Turkey, Poland, and the United States.

In recent years, the top 10 percent of families in Canada took home an average of over 13 times the family income of the lowest 10 percent. This compares with 5.6 times in Finland and is twice as much as the average in the Nordic countries. In 2004, the gap between rich and poor in Canada was greater than at any time in the previous 30 years, and when the figures for 2007 are in it will almost certainly be by far the greatest in modern Canadian history.

How do most Canadians feel about all of this, the rich getting richer and richer, middle class incomes stagnating, and the income of the poor either standing still or declining? In late 2005, a Pollara public opinion poll summed it up nicely as reported under the *Globe and Mail* header "Canadians Not Seeing Benefits of Growth": "While a record number of Canadian believe the nation is in a period of strong economic growth, the majority say they're still not seeing any personal benefits. . . . Only 11 percent of Canadians believe their household income this year will more than keep pace with the cost of living."[8]

A fall 2006 Environics poll said that a record 76 percent of Canadians believed the gap between rich and poor had widened over the past decade. (By the way, British Columbia, the province with the highest child poverty rate in Canada is also the province with the biggest income gap between the rich and the poor.)

A November 2006 report by the Canadian Centre for Policy Alternatives showed that Canadians believed

> that if the rich keep getting richer and the poor keep getting poorer, Canada will end up being more like the United States.
>
> 65 percent of those polled believed that most of the benefit from Canada's recent economic growth has gone to the richest Canadians and hasn't benefited most Canadians.[9]

Another CCPA report, *The Rich and the Rest of Us,* this one by economist Armine Yalnizyan, showed that 40 percent of Ontario families with children (over 600,000 households) saw little or no gain in their incomes for the past 30 years.[10] Thomas Walkom, writing about the report in the *Toronto Star,* put it this way: "There is a party going on, but most of us aren't invited."[11]

When you consider what you've just read on the distribution of income, perhaps you will be interested in a May 2007 *Globe and Mail* editorial which had this to say: "It is a mark of a healthy society when incomes grow and no one is left behind. The system is working."

Amazing!

THE DISTRIBUTION OF WEALTH IN CANADA

THE RICH ARE GETTING RICHER,
AND RICHER, AND RICHER

L et's move on from wages and income to look at wealth in Canada. The figures that follow are up to 2005, but there's lots of evidence to suggest that when more recent breakdowns are available, the imbalances below will be shown to have worsened considerably.

Statistics Canada has reported that between 1984 and 2005 only the wealthiest 10 percent of Canadians increased their share of the national net worth, while the other 90 percent of families saw a decline in their share of wealth, meaning all assets, such as houses, property, cars, bank accounts, shares, and so on. While the bottom 10 percent of Canadian families saw their wealth fall between 1984 and 2005 by an average of $7,500, during the same years the top 10 percent of families more than doubled their wealth from $535,000 to $1,194,000.[1]

By 2005, the highest 20 percent of Canadian families owned 69.2 percent of all net worth. The next highest 20 percent owned 20.2 percent. So, the top 40 percent owned 89.4 percent of Canadian net worth. At the other end of the scale, the bottom 60 percent of Canadians owned only 10.6 percent of net worth, and the poorest 40 percent owned only 2.4 percent.[2] That's right. The bottom 40 percent of Canadian families owned only 2.4 percent of all family wealth in Canada!

Measured another way, the top 50 percent of individual Canadians owned 96.2 percent of all wealth, and the bottom 50 percent controlled only a tiny 3.8 percent.

Whichever way you want to measure it, the figures are pretty appalling. And the disparities are even greater since Statistics Canada does not include the value of registered pension plans in their wealth calculations.

Taking another look at the 1984–2005 comparisons, between 1999 and 2005 the median net worth of the top 20 percent of Canadian families increased by 19 percent, while the lowest 20 percent saw their net worth remain essentially unchanged.

The annual Statistics Canada report on Canada's national net worth is inevitably followed by cross-Canada press reports such as "National net worth reached $5.3 trillion . . . or $159,900 per person."[3] Such reports are, of course, almost meaningless when distributions of both income and wealth are so warped. A Canadian Imperial Bank of Commerce economist noted that despite "one of the strongest economic expansions in many years . . . we still haven't seen a closing of the gap (between rich and poor). In fact we've seen a widening."[4]

The research and advisory firm Economist Intelligence Unit says that the super rich in Canada (people worth $3-million or more) will go from 90,000 households in 2006 to 320,000 by 2016, while the number of individuals whose wealth is over $1-million will increase sixfold, to 2,100,000, a rate of increase double that of the United Kingdom or Germany.

Canadian Business magazine says there are now 46 billionaires in Canada, whose combined net worth is greater than the total assets of the bottom 14 million Canadians.

Let's try that again. Forty-six individuals with greater total wealth than 14 million of their fellow citizens!

Whichever way you choose to look at it, during the past two decades in Canada, the rich have been getting much richer, the poor have been getting comparatively poorer, and most hard-working Canadian families and individuals still aren't being invited to the party.

If you get a chance, have a look at the 2007 *Canada Year Book*. It says Statistics Canada analyzed 13,348 family units. The top 5.7 percent in 2005 owned over 97 percent of all family wealth. The bottom 94.3 percent together owned less than that 3 percent.

Sound fair to you?

BIG-BUSINESS BELLYACHING AND
ACTUAL CORPORATE PROFITS IN CANADA

"We have now found who's claiming
the lion's share of gains in the economy."

Now let's look at how corporations in Canada have been doing. Let's start by looking at net profits. In 2003, they reached an all-time record of over $102.6-billion. The next year, they increased to over $132.3-billion. In 2005, they jumped to over $157.5-billion. And in 2006, there was yet again another new record – of over $168.2-billion in net corporate profits.[1] In the third quarter of 2007, they set another new record of $67-billion. In 2003, 2004, 2005, and 2006, net corporate profits in Canada totalled $560.75-billion, far greater than in any previous four-year period in Canadian history. In 2006, total corporate earnings of the top 1,000 firms were up 24 percent over the previous year.[2]

In 1992, corporate profits before taxes were 4.7 percent of GDP. In 10 of the next 14 years, they increased as a percentage of GDP. By 2005, they were up to 13.8 percent of GDP, and in 2006 they were up again, to 13.9 percent of GDP. In the words of Global Insights economist Dale Orr, "We have now found who's claiming the lion's share of gains in the economy since 1998." And, as most of us will have suspected, all of the top 10 most profitable Canadian corporations in 2006 were either oil and gas companies or banks and other financial institutions.

As corporate profits increased, wages declined as a percentage of GDP. As we've seen, in 1992 they were 55.4 percent of GDP. By 2005, they had fallen to only 50.2 percent. That 5.2 percent difference represented a decline of an enormous $71.3-billion in the share of GDP represented by wages in 2005. That's a huge amount of money missing from pay-cheques. Yet if you listen to all the constant corporate and lobby-group bellyaching from big business and the organizations they fund, such as the C.D. Howe Institute, the Fraser Institute, the Canadian Council of Chief Executives, the Canadian Chamber of Commerce, etc., etc., the picture you get is that business in Canada is hard done by, overtaxed and over-regulated, and conditions here are lousy for investment. As detailed in the chapters that follow, all of this is nonsense.

Let's have a quick look at U.S. after-tax corporate profits and compare them with Canadian corporate profits. In 2005, U.S. profits were 7.5 percent of GDP. In the second quarter of 2006, they reached 8.6 percent of GDP, the highest level in 60 years! In Canada in 2006, they were a huge 11.63 percent of GDP. Surprised? Well, from 1982 to 2006, net profits in Canada were higher than those in the United States as a per-centage of GDP in 16 of those 24 years! Poor, hard-done-by, overtaxed corporate Canada!

Keep this in mind next time you read yet another *National Post* article by someone like Jack Mintz about how we must have even more big tax cuts for corporations in this country.

There are other ways of comparing corporate profits. Statistics Canada compares the ratio of operating profit to operating revenue. In 2005, it was the highest since the agency began to make such comparisons.

Because of corporations' huge profits, the ratio of corporate debt as a percentage of equity in Canada fell sharply, from over 70 percent in 2002 to 50 percent in 2006. From 2000 to 2005, business in Canada reduced its borrowing by an enormous $276.2-billion.

In 2005, the following sectors of the Canadian economy all had all-time record operating profits: primary industry, including oil and gas; transportation; finance and commerce; insurance; and real estate. Oil and gas companies in Canada, mostly foreign-owned, increased their profits by $10.5-billion in 2005, more than 47 percent over their 2004 results. In 2006, EnCana posted the biggest annual corporate profit in Canadian history, some $6.4-billion. The same year, the primary oil and gas industry in Canada had an all-time record operating profit of $39.9-billion.

All told, foreign corporations took almost 31 percent of 2005 profits and an estimated 34.1 percent of 2006 profits. In 2005, over $22.3-billion of foreign-controlled corporate profits left Canada, mostly for the United States.

Let's not forget Canada's banks. In fiscal 2007, their after-tax profits were a huge $19.5-billion.

In the next part of this book we'll look at what corporate Canada is doing with its profits. It's not a pretty picture.

PART FOUR

BIG-BUSINESS INVESTMENT IN CANADA

THE WEAKEST IN OUR HISTORY
COMPARED TO RECORD PROFITS

Among the many lavish promises made by big business's Business Council on National Issues (BCNI), now known as the Canadian Council of Chief Executives (CCCE), was that with guaranteed access to the U.S. market because of the FTA, Canada would surely become a secure base that would attract companies from around the world, and business investment in new machinery and equipment would explode. During the first 17 years of the FTA (from 1989 to 2005), however, what actually happened was that the rate of investment to GDP was almost identical to the previous 17 years, and more recently, in the period from July 2006 to July 2007, equipment imports actually fell, despite the huge increase in the value of the Canadian dollar. The growth of machinery and equipment capital stock in Canada has been a pathetic 0.3 percent during the last five years compared to 4.6 percent during the previous five-year period.[1]

Given the huge corporate profit increases, one would naturally have expected that there would also have been large increases in domestic business investment as a share of the economy. But this hasn't happened. Let's compare corporate profits to the extremely important business investment in machinery and equipment. In 1998, business investment in new machinery and equipment amounted to 86 percent of profits after taxes. After years of record corporate profits, in 2006 it fell all the way down to only 47 percent. Measured another way, business investment in new machinery and equipment was 8.1 percent of GDP in 1998 and 1999, but in 2006 it was down to 6.5 percent of GDP. Business investment as a component of final domestic demand has fallen from a high of some 14 percent in late 2005 down to under 5 percent in late 2007.

So, as we've just seen in the preceding chapter, while there have been all-time record high corporate profits, far greater than in all other years in 2003, 2004, 2005, and 2006, business investment in machinery and equipment, instead of increasing, fell dramatically to new lows. Looking at real business investment in new machinery and equipment, taking inflation into consideration, with 2000 as a base year equalling 100, the number has fallen every year since, all the way down to only 83.5 in 2006.

Of course, all of this raises the question of why we should pay any attention at all to the apparently never-ending corporate clamour for even lower corporate taxes. The Mulroney, Chrétien, Martin, and Harper governments have all lowered corporate tax substantially, yet instead of investing a greater share of their burgeoning profits in new and better machinery and equipment to help increase productivity, corporations in Canada reduced investment to new contemporary lows.

Let's measure business investment in new machinery and equipment another way. As corporate profits increased, business investment in machinery and equipment declined as a percentage of GDP. In both 1998 and 1999, these investments represented 8.1 percent of GDP. Since then, they declined over the next three years to a low of 7.0 percent and then declined even further to only 6.6 percent of GDP for the next three years. In 2006, they were down to only 6.5 percent. The decline of 1.6 percent of GDP since 1999 represented lower machinery and equipment investment of a huge $23.1-billion in 2006.

One would have thought that the sharply increased value of the Canadian dollar would have encouraged large increases in investment in new imported machinery and equipment. David Crane of the *Toronto Star* explains: "When our dollar was at 65 cents, a $2 million (U.S.) piece of machinery would cost $3.1 million (Canadian); when our dollar is at 95 cents, the same technology costs $2.1 million (Canadian) or one-third less."[2] But even with a high dollar and record profits, the rate of real investment growth fell from over 25 percent in 1998 to only 10 percent in 2005. Meanwhile, hourly productivity growth in the business sector, as we shall see, has been very weak. Statistics Canada summed things up: "Over much of the last decade, corporations have been posting record profits. Meanwhile, business fixed capital investment has been relatively sluggish in recent years."[3]

In a February 2006 *Globe and Mail* column, economist Jim Stanford wrote:

> Canadian corporations are raking in more money than at any time in history. And they aren't spending it on what our economy needs – in this case productivity-enhancing invest-ments in technology and equipment. The corporate sector has amassed a hoard of cash. . . . [There has been] a 50 percent surge in after-tax cash flow since 2000. Canadian businesses currently sit on $280-billion worth of cash, foreign currency and short-term paper.

And all this while large dividends and inter-corporate fees have flowed out of the country, with much of it headed for tax havens.

It's interesting to compare investment by business in Canada and the United States as a percentage of GDP. The Ottawa-based Centre for the Study of Living Standards had this to say in their fall 2005 *International Productivity Monitor* under the heading "Under-Investment in Capital":

> Between 1991 and 2003, Canada's private sector invested about 13 percent less per dollar of GDP in machinery, equip-ment and software than their counterparts in the United States. This under-investment slowly eroded the relative strength of our capital stock. This erosion in turn reduces the productivity of our labour force and hence our prosperity.
>
> Our under-investment is a major factor in explaining the $7,200 GDP per capita or 15.7 percent shortfall between us and the United States.

Buzz Hargrove, president of the Canadian Auto Workers, had this to say: "Measured as a share of GDP, new business investment spending (excluding the booming energy sector) has been weaker since 2001 than during the last recession. And measured as a share of corporate cash flow, it is the weakest in our history."

Aside from the above criticisms, somehow very few commentators have tackled our business leaders' failure to invest in our own country while they pump billions of dollars out of Canada every year.

The OECD is to the point: "Non-residential investment has decelerated. . . . In particular investment in machinery and equipment has slowed sharply."[4] In fact, in 2006 there was actually a decline in manufacturing investment.

Even the usually pro-big-business TD Bank Financial Group chief economist Don Drummond, a former federal associate deputy minister of finance, says Canada's business leaders have not "fully responded to the investment opportunities offered by strong profit growth and declining prices of imports of machinery and equipment due to the stronger Canadian dollar."[5] And James Milway of the Institute for Competitiveness and Prosperity describes the "lethargy" among business leaders: "Our businesses are not investing enough in machinery and that is why we are not as competitive."[6]

To repeat: all this despite the combination of much higher corporate profits, much lower corporate taxes, and a substantially strengthened Canadian dollar. In November 2005, Ottawa's *Economic and Fiscal Update* said, "Corporate profits, at 14 percent of GDP, are at the highest level in 30 years." The JP Morgan Chase chief Canadian economist said, "Corporate earnings and corporate balance sheets are in their best shape on record . . . earnings are at a record level to GDP."[7] Contrast this with a Statistics Canada study that describes business investment as "sluggish," noting, "There has been slower capital expenditure. In particular, capital expenditure on machinery and equipment has accounted for more of the weakness."[8]

How does Canada compare with other countries aside from the United States? Looking at the most recent three-year period for which comparative international statistics are available, 2003 to 2005, Canada was way down in 18th place in gross fixed capital formation (fixed assets which will be used in production for several years) and in investment in machinery and equipment as a percentage of GDP. All of the following countries had a better record than Canada. In order, China, Korea, Spain, the Czech Republic, Australia, the Slovak Republic, Ireland, Greece, Iceland, New Zealand, Japan, Hungary, Portugal, Switzerland, Austria, Luxembourg, and Italy.[9]

On August 14, 2007, the TD's Don Drummond and economist Ritu Sapra produced a "special report" titled *Canadian Companies Not Taking Advantage of Investment Opportunities.* As a result of companies' failure to invest,

> Canada's productivity growth record has been dismal, both from a historical and an international perspective.
>
> One of the main causes is our lagging machinery and equipment investment, both in absolute terms and relative to other countries. And yet, the investment climate of recent years could scarcely be any more positive.
>
> Corporate profits have soared to their highest share of GDP since at least the early 1960s.
>
> The past decade has seen a declining trend in business sector investment intensity in Canada compared to other OECD and G7 countries. The comparison with the U.S. is even worse.
>
> To illustrate, for every dollar of new investment enjoyed by the typical OECD worker in 2006, his or her Canadian counterpart got only 87 cents. And for every dollar per employee spent on new investment in the United States, only 75 cents were spent in Canada.
>
> Put another way, in 2006, the average Canadian worker received some $594 less in investment spending than the typical worker in OECD countries a good $1131 less than the average G7 worker.

The Canada-U.S. investment gap is almost entirely the result of lagging machinery and equipment investment.

In 2006, investment in new machinery and equipment as a share of profits was at an all-time low. One wonders when enough will be enough for big business and their never-ending demands for even lower taxes. Is there no limit to corporate greed?

Incredibly, after record-breaking profits yet again in 2006, new corporate investment in Canada was forecast to grow only by a tiny 1.9 percent in 2007.

RESEARCH AND DEVELOPMENT

ANOTHER BIG-BUSINESS FAILURE

Here's a warning. If you read this chapter you may find it very difficult to suppress a laugh the next time you go to a chamber of commerce or Rotary meeting and hear someone from the Canadian Council of Chief Executives (CCCE) or the C.D. Howe Institute give a scary speech about low productivity and the inevitable resulting fall in our standard of living, plus the crucial need for increasing innovation and more tax cuts in Canada.

In the next chapter, we'll turn our attention to the important question of productivity. But first, we'll look at the dismal record of big business – despite huge corporate profits and among the very best tax incentives in the world – to do anywhere near the amount of research and development necessary to spur productivity.

Let's begin by comparing R&D expenditures in Canada to R&D in other OECD countries as a percentage of GDP. In the period from 1997 to 2002, Canada was down in 15th place at 1.9 percent. Israel headed the list at 5.1 percent, followed by Sweden at 4.3 percent, the Netherlands at 3.5 percent, and then the United States and Iceland at 3.1 percent.[1]

The 2006 United Nations *Human Development Report* list of R&D expenditures for the period from 2000 to 2003 also placed Canada 15th on the list, again at only 1.9 percent of GDP. But in another comparison, this one from the OECD, Canada is way down in 25th place in the percentage of R&D performed by industry.

Some further comparisons: in Luxembourg, business accounts for just under 90 percent of all R&D. Canada's industrial companies' R&D of roughly 47 percent was below the G7 average (62.75 percent), the EU average (53.66 percent), and the total OECD average (61.93 percent). Canada's business R&D is far below that of countries such as Korea, Japan, Switzerland, Sweden, Belgium, Finland, Germany, and Denmark. Other countries that also have a higher percentage of business-sector R&D than Canada are Australia, Greece, Ireland, Mexico, Portugal, Spain, and Turkey.

In sharp contrast, Canada leads all OECD nations in higher education R&D expenditures as a percentage of GDP. Only Finland, in second place, comes close.

However, business R&D as a percentage of GDP in Canada is well under half the rate of the United States, and only about one-third the rate for high-tax Sweden, while the federal government's R&D contribution is below the OECD average, and is expected to be lower still in 2008.[2]

Want to know where the basic problem lies? The 2006 OECD *Economic Survey of Canada* puts it gently: "Federal and provincial governments have a more generous array of tax credits and grant programmes designed to encourage business R&D expenditures than most OECD members. Nevertheless, business expenditures on R&D as a share of GDP remain lower than in many OECD countries."

As giant new annual corporate profit records have been set, corporate R&D spending in Canada as a percentage of revenue has, since 2001, remarkably been declining instead of increasing. Incredibly, the highly profitable oil and gas sector spent a pathetic 0.36 percent of corporate revenues on R&D in 2006, while the industrial average for corporations in Canada was 3.5 percent.

A good reflection of how poorly we do in R&D can be found in patents. In a list of countries, Canada was in 30th place in patents granted per million people during the years 2000 to 2005, far below the OECD and European Union averages. Measuring the number of patent applications per billion dollars of GDP, Finland, Japan, and Switzerland were all at four or over, Sweden and Israel were over three, while Canada was at fewer than one per billion dollars of GDP. Measured another way, by the number of patents filed by Canadian residents per $1-million of R&D spending, Canada is ranked 26th.[3] Yet another way to measure patents granted is by the number granted per million people. Here, the Canadian record is just awful, at 35 in 2004. This compares with 874 for Japan, 738 for Korea, 281 for the United States, 275 for Sweden, and 222 for Finland.

In the information and communications technology (ICT) index, which measures per-capita usage of telephones, the Internet, personal computers, and mobile phones, Canada is down in 16th place. When it comes to the percentage of households with a home computer, in 2004 Canada was in eighth place among OECD nations, at about 69 percent, far behind Iceland at almost 86 percent, Denmark at over 79 percent, and Korea and Japan at just under 78 percent, but well ahead of 17 other OECD countries, including Australia, the United Kingdom, New Zealand, and also the United States (just over 65 percent).

The percentage of households with access to the Internet also varies substantially in OECD nations. Heading the list is Korea at 86 percent, followed by Iceland at just under 81 percent. Canada, at just under 70 percent, is just above the level of the United States, but well ahead of many other countries, including Ireland and Italy (both at 39.7 percent), Austria, the Czech Republic (19.4 percent), France, Greece (16.5 percent), and Hungary (14.2 percent). Japan was just under the Canadian level in 2004, as was the United Kingdom, at 56 percent.

When it comes to the share of ICT equipment in total manufacturing value added, Canada is in 15th place, far, far behind such countries as Finland, China, the United States, Japan, Germany, Korea, the Netherlands, the United Kingdom, and even Mexico, but ahead of 14 other OECD countries plus Russia, Brazil, and South Africa.[4]

If we look at high-technology exports as a percentage of total manufacturing exports – that is, the exports of aircraft, computers, pharmaceuticals, motor vehicles, electrical equipment, and so on – Canada is way down in 39th place. We're well below the OECD and EU averages, and even below Mexico. Despite the perpetual corporate rhetoric about our need for innovation so we can be more competitive, our high-technology exports as a percentage of all our exports were lower in 2003 than they were in 1991, and well below the levels for 2000 and 2001.

Some comparisons. The United States' high-tech exports as a percentage of total manufacturing exports amounted to 35.8 percent in 2003. The United Kingdom was at 34.7 percent. The OECD average was 24.5 percent, the EU 22.1 percent. Canada's high-tech exports as a percentage of total manufacturing exports were a very poor 12 percent.

Measuring our overall manufacturing exports as a percentage of all exports leaves Canada in a dismal 34th place. But what else can you really expect when big business in Canada prefers to invest out of the country or keep its money in the bank or in foreign tax havens instead of investing to help build a more competitive, innovative, and productive country?

In spring 2007, Statistics Canada said it well. For most OECD countries, "growth in labour productivity was highly responsive to business research and development.[5]

In the next chapter we'll see how corporate Canada's failure to turn its record profits into meaningful investment is affecting the standard of living of all Canadians.

PRODUCTIVITY IN CANADA

"Business in Canada gets a low mark . . .
woefully short of international results."

To some, productivity is almost everything. Here's Jeffrey Simpson of the *Globe and Mail:* "A politician who wanted to talk straight to the Canadian people . . . would discuss productivity morning, noon and night."[1] In another column, Simpson writes, "Without productivity growth, there's no real economic growth, no real wealth creation, no improvement in the country's overall standard of living."[2]

Not many economists would disagree. But now Simpson quotes from a study by a Toronto consulting company, Impact Group, which hits the nail squarely on the head, though it's not the sort of thing you'll ever hear from the Canadian Chamber of Commerce or the Canadian Council of Chief Executives: "Very few Canadian companies do research and development. Over seven years, only 9 percent of Canadian firms did R&D every year. The remainder did nothing or were 'at best occasional performers.'"

Then comes something I never thought I'd ever see from Jeffrey Simpson: "The impact study doesn't say precisely why [there's so little R&D], but here's a guess. The hidden issue of Canadian economic structure is foreign ownership. Canada remains too much a branch-plant economy, and branch plants don't do much R&D unless bribed to do so by governments."

Well yes, but a "hidden issue"? Some of us have been speaking and writing about this hidden issue for decades. The manufacturing industry, which in other countries does much of business R&D, is over 50 percent foreign-owned in Canada. Why would big transnational corporations, mostly American, transfer their R&D out of their own countries to Canada, even with our big government incentives? And for U.S. companies specifically, why would they even consider doing so, given their government's post-9/11 security concerns and paranoia.

According to our corporate community, our poor performance in productivity can be chalked up largely to our "high" taxes. (We'll learn more about that mythology shortly.) But could there be another explanation? Philip Cross, Statistics Canada's highly regarded chief of current analysis, writes, "As a long-time student of business cycles stated, 'Of all the components of aggregate demand, it is the investment spending by firms and households that is the prime mover in economic fluctuations.' . . . This fact has long been known and recognized."[3]

Right, and we've just seen in the last couple of chapters exactly how very poorly corporate Canada measures up in this respect.

How poor is our productivity? According to the OECD, in a list of 24 member nations, Canada was fifth in productivity in 1970, 12th in 1980, 16th in 1990, and 16th in 2000, and between 2000 and 2005 we were 20th of 29 nations in GDP per hours worked.[4] By 2006, we had fallen further still, to 22nd place and by 2007 we were 47th in a list of 50 developed countries. Nice little trend.

Canada's poor productivity performance compared to that of the United States should be viewed in relation to the extraordinarily lavish promises about the certain gains to be made following the signing of the FTA and NAFTA. As with so many other misleading or inflated promises, it hasn't exactly worked out as we were told it would. Follow the figures.

Between 1960 and 1988, Canada's GDP per worker was over 90 percent of the U.S. level for 24 of those 28 years, with a high of over 94 percent. In 1988, the year before the FTA came into effect, GDP per worker in Canada was 90.95 percent of the U.S. level. By 2005, after 17 years of free trade with the United States, it was all the way down to only 84.4 percent of the U.S. level.[5] By 2006, it was down again, to 82.6 percent. In the first 17 years of the FTA, from 1989 to 2005, the annual growth rate in output per worker in Canada was lower than that for American workers in 13 of those 17 years.

Focusing specifically on business sector productivity, GDP per hour in Canada in 1988 was 86.21 percent of the U.S. level. By 2005, it was down to only 74.41 percent, the lowest level in 51 years! In 2006, we were down again, to 73.95 percent. Between 2000 and 2005, business sector productivity in Canada grew by only 1 percent compared to 3.3 percent in the United States.

In the period from 1995 to 2004, the average annual growth in GDP per hour worked in the OECD was 2.2 percent. In Canada, it was only 1.7 percent. All of the following countries were above 2 percent: Australia, Ireland, the United Kingdom, the United States, Greece, France, Finland, Norway, and Sweden.

Even some of Canada's foremost economists seem perplexed by the extremely weak productivity growth in Canada. Terms like "an unprecedented divergence" and "a mystery" have not been uncommon. The developments are frequently referred to as "troubling" and "threatening to Canada's future prosperity."

Productivity expert Andrew Sharpe of the Centre for the Study of Living Standards points out that

> most of the obstacles identified by the business community in the 1980s and 1990s as impeding productivity growth have been removed, yet productivity growth has never been worse.
>
> The obvious question is why.
>
> Growth in the machinery and equipment capital stock and capital-labour ratio has fallen off in recent years, reflecting slower investment growth. This has reduced the rate at which new productivity enhancing technologies are put into operation.

Even the *Globe and Mail*'s editorial page describes the private sector's investment in the adoption of new technologies as "pitiful": "Many experts have blamed the failure to invest in information and communications technology, in particular, for more than half the productivity gap with the United States."[6]

David Crane of the *Toronto Star* also has it right:

Productivity is about innovation – about skilled employees working with advanced technologies generating high-value products and services that support good jobs and create the wealth for a successful society. Productivity is not about a race to the bottom, relying on low wages and a cheap dollar for success in the global economy. That is the route to a failed society.

And, to emphasize his point:

One problem in Canada is that businesses are much slower to equip their employees with the latest technologies than U.S. companies.

Innovation is critical in enhancing productivity. Here . . . business in Canada gets a low mark.

In 2005, the Information Technology Association of Canada said that Canadian companies invest only 43 percent of what U.S. companies spend per worker in information and communications technology. For 2003, it was $1,384 (U.S.) in Canada compared to $3,235 (U.S.) in the United States.

Let's focus on the foreign-dominated manufacturing sector in Canada. Using 1992 as a comparative base year equalling 100, our output per hour in manufacturing in 2003 was 134.5 compared to 180.4 for the United States and 154.3 for Japan. In 1980, our manufacturing productivity gap with the United States was about 13 percent. By 1989, when the FTA kicked in, it was about 18 percent. By 2001, it had increased to a huge 33 percent. In 2000, labour productivity growth was over 6 percent. By 2006, it had fallen to minus 2 percent.

In January 2006, a *Globe and Mail* editorial on the productivity gap lamented Canada's terrible record in productivity growth and the negative consequences for our future standard of living. Blaming "the private sector's pitiful investment in the adoption of new technologies" and the failure to invest in information and communications technologies, the *Globe* advocated yet another round of tax breaks for large corporations.

But wait a minute. Big business has had a dismal record in new investment, not only new technologies but also overall in badly needed new machinery and equipment. This, despite the fact that in recent years corporate net profits have been at all-time record highs.

So if big business has been making record profits year after year and still not investing anywhere near enough to make us competitive in productivity, why advocate more tax breaks for the same gang? It makes no sense, particularly when corporate Canada's share of national income has been increasing steadily while most Canadians have had to get by with tiny, if any, growth in their incomes. Based on their track record, big business (much of it foreign-owned) doesn't deserve more tax breaks.

In 2005, at 2.1 percent productivity growth, Canada finally rebounded from the truly dismal record of 2000 to 2004, but by 2006 we were back down at only 1.2 percent. For 2007, estimates were that productivity growth in Canada would again be far below that of Finland, Japan, Germany, Britain, and France, and once again well below the U.S. level.

It's interesting to note that in the total service sector of the economy in recent years, Canada has outperformed the United States in growth rates. This includes communications, the retail trade, gas utilities, and business services.[7] Could one of the reasons for this be that the service sector of the Canadian economy is mostly Canadian-owned and Canadian-controlled?

As we did when we began this chapter, let's again consider the question of the relative importance of productivity. TD economist Don Drummond wrote in the fall 2006 issue of the *International Productivity Monitor,* "Despite poor productivity growth, Canada remains a wealthy country. But there is ample reason for concern. Unless our record is turned around quickly Canadians' quality of life will stand still while other nations move ahead. The recent Canadian record falls woefully short of international results." A headline in the *Financial Post* warns, "Productivity Slump Could Take Toll on Living Standards."

In the aforementioned study by Don Drummond and Ritu Sapra, after documenting the poor investment record by Canadian business, the two authors focus on the inevitable results for productivity.

> Productivity growth has slowed dramatically over the past several years – a development that threatens the well-being of Canadians.
>
> Looking ahead, productivity will be an increasingly important determinant of economic growth, especially in the face of the demographic crunch that is looming in Canada's future.
>
> Canada's abysmal productivity performance has resulted in the widening of the business sector labour productivity gap versus the United States.
>
> Since 1973, Canada has had the third lowest rate of growth in output per hour among 23 OECD countries. This resulted in Canada's level of productivity falling from the third highest in the OECD in 1973 to 16th in 2006.

So can you blame poor investment and poor productivity on our lack of business profits? Hardly. Can you blame a too-low Canadian dollar? Hardly. Can you blame corporate taxes, as our constant corporate whiners have been doing daily? Hardly.

Drummond and Sapra point out that corporate taxes in Canada have had little effect on the corporate sector's after-tax profitability. "Also, the marginal effective tax rate on business investment was slightly lower in Canada than in the United States . . . the corporate tax system can't be blamed for Canada's investment shortfall, compared to the United States."

In fact, as we shall see shortly, Canada's corporate taxes are now well below the level of corporate taxes in the United States.

Late in December 2007, Statistics Canada reported that R&D growth in the business sector in 2007 was expected to have grown by $416 million. This amounts to less than one quarter of one percent of 2006 net corporate profits.

16

MANUFACTURING IN CANADA

"There's blood on factory floors."

The FTA, NAFTA, globalization, and our much higher dollar have all taken their toll on manufacturing in Canada. But poor levels of R&D, inadequate investment in machinery and equipment, and, of course, competition from increasingly competitive low-wage countries have all been factors.

Employment in manufacturing in Canada peaked in November 2002, but by early 2008 it was down by a huge 348,000 jobs.[1] This enormous decline is very significant, because manufacturing jobs are mostly high-paid with good benefits and usually have a big multiplier impact in other areas of the economy. CIBC economist Jeff Rubin has predicted that by the end of the decade job losses in manufacturing might be over 500,000.

The manufacturing industries in all of the following countries contribute a higher share of total gross value-added in their countries than manufacturing does in Canada: the Czech Republic, Finland, Germany, Hungary, Ireland, Italy, Japan, Korea, the Slovak Republic, Sweden, and Switzerland.

In a list of the top 40 countries back in 2005, measured in dollar value, Canada had the eighth largest manufacturing output, behind the United States, Japan, China, Germany, the United Kingdom, Italy, and France, but ahead of such countries as South Korea, Brazil, Russia, Spain, India, and Sweden. In terms of overall industrial output, we were also in eighth place. Note that all the countries ahead of us in these rankings have considerably larger populations than Canada. But note, too, the devastating decline in manufacturing in the country during the last few years.

Back in 1970, manufacturing jobs accounted for about 22 percent of all jobs in Canada. Now, it's less than 12.5 percent. While it's true that the share of manufacturing jobs has fallen in all G7 countries, in Germany and Italy it is still over 20 percent, and Japan and France are both ahead of Canada in this respect. Notably, though, the U.S. manufacturing job share has fallen from about 25 percent in 1970 to only 9 percent. The United Kingdom's rate fell from 35 percent to some 14 percent.[2]

Canada has more of its manufacturing industry under foreign control than any other OECD country, over 50.3 percent by 2004, and certainly much higher than that today. Only Ireland, Luxembourg, and Hungary are close. All the other OECD countries have their manufacturing between 5 percent foreign-owned (Germany) and 32 percent (Sweden). In the United States, it's only 12 percent.

In 1970, manufacturing accounted for about 23 percent of Canada's GDP. In 2007, it was down to about 15 percent. Current manufacturing employment in Canada is at its lowest level in 10 years, and its share of total employment is at its lowest level since the end of the Second World War. By 2006, Canada had fallen to 25th place in the list of countries exporting manufactured goods, and was down to 65th in terms of manufactured goods as a percentage of all merchandise exports.

There are a few who say, "What's the worry? National unemployment has been at its lowest level in three decades." Among the problems with that logic is the fact that a very high percentage of new jobs have been low-paid, part-time, and/or self-employed positions, instead of high-paid, full-time, secure jobs with good benefits.

As a leader in the *Globe and Mail* pronounced, "There's Blood on Factory Floors."[3]

In October 2007, the research and policy firm Informetrica said that Canada's increasing trade deficit in manufactured goods, some $28-billion in 2006, is unsustainable.

17

CARS, TRUCKS, AND AUTO PARTS

THE GOOD NEWS AND SOME QUITE BAD NEWS

The good news is that in 2007, Canada had its second best year ever in sales by new vehicle dealers (1.66 million, just below the 1.73 million record in 2002).[1] The bad news is that in 2006 Canada had its first auto trade deficit in 18 years, a deficit of just over $1.2-billion.[2] Statistics Canada describes it as a "spectacular change" in our trade balance. The projected automotive trade deficit for 2007 was expected to be a record, a huge $8-billion, with offshore imports representing almost a quarter of all sales.[3] (In 1999, our automotive trade surplus was just under $15-billion.)

Canadian Auto Workers (CAW) President Buzz Hargrove explains: "It's both the imports that are coming into North America, and the lack of ability of our vehicle builders to export to China, Asia and the European Community, who all have their markets essentially closed and protected."[4]

More bad news is that motor vehicle production in Canada has fallen from a high in 1999 of 2,735,257 units to 2,081,487 in 2006, a huge drop of 653,770, and exports dropped from $97.9-billion in 2000 to $74.7-billion in 2006. In August 2007, Statistics Canada reported that the five lowest levels of exports since the end of 1998 were during the previous five quarters. Meanwhile, 2006 was a record year in Canada for the sale of vehicles that were manufactured overseas.

More bad news. In 1993, Canada was the world's fourth largest auto assembler, but by 2005 we had dropped down to eighth place, and we will likely have fallen to tenth by the end of 2007. While this has been happening, Honda and Toyota are continuing to take market share from Ford, General Motors, and Chrysler, while Toyota in early 2007 passed General Motors in global sales for the first time and foreign automakers captured over 50 percent of the U.S. market for the first time.

As the "Big Three" automakers – which directly and indirectly support 300,000 Canadian jobs – lose market share, millions of vehicles come into North America from offshore, while at the same time Honda and Toyota are producing even more cars and trucks on this continent. In 2006, Japanese automakers produced just over 900,000 automobiles in Canada, more than double their 1998 production, and their share of the domestic market reached 34 percent. In Canada, overseas brands have outsold cars made in North America every year since 2001, and the gap is growing.[5]

Many complain about how Asian countries make it difficult to export cars into their countries, but the failure of the Big Three automakers to respond to growing market demands is an even bigger problem than restrictions on sales abroad.

In Canada, productivity had been growing faster in auto production than in other manufacturing, and faster than in other countries. But real wages have grown at less than half the levels of productivity growth. Buzz Hargrove points out that Canadian assembly plants "are about 5% more productive than in the United States, and 25% better than Mexico."[6]

While Canada was expected to have an almost $10-billion automotive trade surplus with the United States in 2007, we will likely have a huge deficit of over $17.5-billion with other countries. As the CAW points out, "For every dollar of automotive products we import from Europe, we sell 13 cents back to them in exports," and "for every dollar we import from Japan, we sell 2 cents back in exports," and "for every dollar we import from Korea, we sell less than half of one cent back." And major, very competitive Chinese auto production is certain to become a factor soon.

While imports account for only about 10 percent of auto sales in

Europe, less than 5 percent in Japan, and less than 1 percent in Korea, imports now make up about 25 percent of all North American auto and truck sales. Economist Jim Stanford has shown that in the past five years Korean automakers sold 470,000 new vehicles in Canada and produced not one in this country. As Stanford has written, "For us, there's no upside from globalization, just the downside."[7] Some estimates put the job losses at 10,000 with 5,000 more jobs in jeopardy. Industry-wide employment is down 15 to 20 percent since peaking in 1991–2001.

Are things going to get better or worse? For years, we have somehow let Asian countries vigorously protect their automakers while we opened up the prosperous Canadian market to them. Incredibly, our federal government was negotiating a free-trade agreement with Korea that would be certain to make matters much worse. Some industry experts suggest that the agreement would increase imports from Korea by between 20,000 and 33,000 units a year.

Moreover, the once-huge automobile assembly cost advantage that Canada had over the United States has been eroded by the big increase in the value of the Canadian dollar and by cuts in health-care benefits for autoworkers in the United States, so that Canadian and U.S. costs are now much closer compared to a one-time $15 (U.S.) an hour advantage.

True, government subsidies of almost half a billion dollars have been a major factor in the decision to locate new plants in this country. But our universal health-care system and our skilled and well-educated workforce are important factors supporting the industry in this country.

While health-care benefits in the United States are being cut back, automakers in that country still spend more on health insurance for their employees than they do on steel. In 2006, the Conference Board of Canada said that health-care and pension costs in the United States added between $1,400 and $1,800 to the price of every new American automobile or truck.

This said, Canadian auto parts manufacturers are being hit hard by growing low-cost competition from China and Mexico that continues to take a larger share of the U.S. market. As some of us forecast before NAFTA was signed, Mexico soon displaced Canada as the biggest supplier of auto parts to the United States, and it has increased its lead over Canada almost every year.

Some things worth noting. About half the content in a vehicle assembled in Canada consists of imported parts, which create lots of jobs, but not as many as might be thought in this country. This said, jobs in the auto industry in Canada are thought to be responsible for 7.5 percent of jobs in the total economy, reflecting a huge multiplier effect.

Next, while the Big Three have been laying off employees here (perhaps as many as 23,000 union jobs), the Japanese automakers have created roughly twice as many non-union, lower-paid jobs in their Canadian plants.

Unfortunately (and many believe unfairly), the World Trade Organization forced Canada to wind up the Canada-U.S. auto pact, which had worked so very well for us, and which the United States had agreed to with little difficulty because we are such a major market for American vehicles. The pact made sure Canada had a good share of North American auto industry investment and employment. But as Jim Stanford has pointed out, the situation has over the past decade been changing rapidly.

> Offshore imports to North America have ballooned 150 percent since 1996, reaching 4.5 million vehicles in 2006, gobbling more than 25 percent of the continental market. . . . Those 4.5 million imports would keep 14 assembly plants running flat out, and create half a million jobs. Meanwhile, North America exported all of 300,000 vehicles to the rest of the world – barely enough to keep one plant running. . . .
>
> North America is the only major automobile market in the world that imports so much more than it exports.

Canada would benefit enormously from a new auto pact. It's just too bad that the Americans produce such comparatively expensive, poorly designed, and inefficient vehicles. And it's really too bad that Sweden, with only just over nine million people, can have developed its own auto industry, but Canada, with almost 33 million people, somehow hasn't been able to do the same. Where Canada once had an automobile and truck surplus as high as $20-billion, at this writing we were headed for an $8-billion deficit in 2007.

The 2007 *Canada Year Book* says "The auto industry is the one that drives the Canadian economy." If so, we're in for more big trouble. Eighty percent of Canadian-made vehicles are shipped to the United States, but estimates indicate that U.S. sales will be down by 400,000 vehicles in 2008.

While we're on the subject of motor vehicles, a little digression. For anyone who has driven in Russia, the fact that Russia has the highest number of road fatalities, with almost 250 a year per million population, will come as no surprise. Poland, with 150, comes next, closely followed by Korea at 147. What is surprising, at least for some, is that the United States then comes next with 145. Canada is in 18th place on the list with 87 per million population. The lowest countries in terms of road fatalities are the Netherlands, with only 49 per million population, Sweden with 53, and Norway with 56.[8] Many of those who have driven in France will be very surprised that their number is only 92.

The number of road fatalities per million population in Canada has been in steady decline, from 150 in 1990, to 95 in 2000, to only 87 in 2004.

CORPORATE TAXES IN CANADA

"Totally out of whack"

H as a week ever gone by in Canada in recent years without a strident cry from big business, the C.D. Howe and Fraser Institutes, the Canadian Taxpayers Federation, and/or the *National Post* that taxes are far too high in Canada and if our country is to be more competitive and productive corporate taxes must be significantly reduced? Big business has long pressured Ottawa and the provinces to lower corporate taxes, and the governments of Brian Mulroney, Jean Chrétien, Paul Martin, and Stephen Harper have obliged. Yet at this writing, Thomas d'Aquino of the Canadian Council of Chief Executives is yet again calling for even more "broadly based tax relief."

In this chapter, among other things, we'll see how taxes in Canada compared with taxes in other countries before the Harper government's massive tax cuts in their October 2007 mini-budget, and how some countries with higher taxes than ours have fared very well. We'll also see how Canadians have repeatedly been badly misled about taxes in Canada compared to taxes in other developed countries.

First, let's compare overall Canadian levels of taxation (before the cuts in the October 2007 mini-budget) with the tax levels in the other developed countries. Most Canadians reading their daily newspaper or listening to the politicians, big-business executives, and open-liners on radio or television are given to believe that we have been badly over-taxed by our federal and provincial governments. But that's just plain nonsense, and it's a shame that more hasn't been done to correct this long-standing fallacy.

In the list of the 30 OECD member countries, Canada was way down in 21st place when total tax revenue was measured as a percentage of GDP. This included all taxes on personal incomes and corporate profits, all taxes on goods and services and on capital gains, all value-added and sales taxes, all social security, payroll, and property taxes, and so on.

Canada's taxes, at 33.5 percent of GDP in 2004, were below the 35.9 percent OECD average, and far below the EU15 average of 39.7 percent. All the following countries were near, at, or above 40 percent. In descending order, Sweden at 50.4 percent, followed by Denmark, Belgium, Finland, Norway, France, Austria, Italy, Luxembourg, Ireland, the Netherlands, Hungary, the Czech Republic, Portugal, Greece, Britain, Germany, New Zealand, Spain, and Poland.[1] Two years later, in 2006, Sweden was still above 50 percent, and all the following countries had total taxes between 40 and 49 percent of GDP: Denmark, Belgium, France, Norway, Finland, Austria, and Italy.

Among the countries with total tax revenue as a percentage of GDP lower than Canada are Mexico and the United States. But how many among us would like to have the same social conditions as these countries, the terribly high levels of abject poverty, the violence, the millions with no health-care insurance, and the lack of other social programs that are expected and normal in most other developed countries?

As we shall see, some of the "high-tax" countries on the above list not only have the best overall social programs in the world, but are also among the most competitive and productive. Note in particular the comments on these two topics in this book, and the ratings for Finland, Sweden, Denmark, and Norway, and the low rates of child poverty in these countries.

All the following countries have higher taxes than Canada as a percentage of GDP, and all have better productivity in terms of GDP per hours worked: Norway, Belgium, France, Germany, Sweden, Denmark, and Britain. And Norway, Belgium, the Netherlands, and France have better GDP-per-hours-worked records than the low-tax United States.

Now that was for 2004. Every year, the OECD publishes *Revenue Statistics*, and some highlights from the section in the 2006 edition titled "Tax Revenue Trends, 1965–2005" follow.

In 2005, total tax revenue in Canada as a percentage of GDP was again down, to 33.5 percent, and again well below the OECD average of 35.9 percent, the EU19 average of 38.8 percent, and the EU15 average of 39.7 percent.

Listening to some of our extreme right-wing commentators in Canada, you would think that tax revenue here has been rising inexorably. Not so. According to the OECD's *Revenue Statistics*, during the past 40 years, the highest that total tax in Canada has been as a percentage of GDP was in 1997 and 1998, at 38.7 percent, above the OECD averages of 36 percent and 36.4 percent. But in recent years, total taxes have been steadily declining. When the numbers are available for 2006 and 2007, the tax-to-GDP percentages will be even lower than 33.5 percent again.

If we look at how taxes changed in the years between 1995 and 2005 as a percentage of GDP, 22 OECD countries increased tax levels, while Canada was one of the 8 countries that decreased taxes. If we consider all-inclusive government revenues, tax and non-tax receipts in 2006, 18 OECD countries have higher revenue as a share of GDP than Canada.[2]

If you compare net financial government liabilities in 2005 as a percentage of GDP, Canada, at 26.3 percent, is far lower than all the other G7 countries. Next is the United States at 38.7 percent, Britain at 40.6 percent, France at 44 percent, Germany at 58.4 percent, Japan at 86.3 percent, and Italy at 98.6 percent.

It's worth noting that the provinces take about 38 percent of the tax revenue in Canada, higher than provinces or states in any other industrialized country.

As in so many other important issues in Canada, in tax policy the gap between the wishes of big business and the priorities of most Canadians is enormous. With rapidly growing corporate profits, assets, and concentration of ownership, the gap has been getting bigger, and after the Harper government's 2007 mini-budget, the gap will be much larger in the future.

As we all know, big business stridently believes large tax cuts for corporations have been an urgently required priority. How do most Canadians feel? A late-2005 poll by Decima Research placed tax cuts in 16th place on a list of Canadians' priorities. And half of those polled said that they didn't feel tax cuts would promote investment or increased productivity. Perceptive people. As you will have already seen in the investment and productivity chapters in this book, large corporate tax cuts in the past failed to produce the promised results. Moreover, numerous studies have consistently shown that factors other than tax rates, including interest rates, land and energy costs, transportation infrastructure, and a well-trained workforce, are more important determinants in investment decisions.

In a *Globe and Mail*/CTV poll just before the 2007 budget, when asked what the most important thing the new budget should address, 50 percent of respondents said increased spending on social programs, while only 19 percent opted for cutting taxes. Improving health care and the environment and reducing child poverty are far more important to most Canadians. So are more affordable post-secondary education, more R&D, and reducing inter-provincial barriers.

Okay, let's have a look at comparative corporate taxes. First, a few samples of the steady stream of misinformation, exaggeration, and just plain BS that Canadians are fed on a regular basis. According to James Milway of the Toronto-based Institute for Competitiveness and Prosperity, "We tax businesses at higher rates than almost any other country in the world."[3] *Globe and Mail* columnist Neil Reynolds writes about Canada's "high corporate tax rates, one of the great economic absurdities of our times."[4] And the *National Post*'s town crier for lower corporate taxes, Jack Mintz, former head of the continentalist C.D. Howe Institute, says that Canada has "one of the highest corporate income tax rates in the world."[5]

The *Financial Post* told its readers on May 18, 2007, that personal and business taxes "are generally higher in Canada than the rest of the industrialized world." *Maclean's,* one month later, told its readers that the rich in Canada have to pay taxes between 39 percent of their annual income to as high as 48.2 percent, somehow confusing the top marginal rates with the much, much lower effective rate of overall taxes actually paid. The same month, the *Financial Post*'s Diane Francis claimed in her column that "Canada has the highest corporate taxes in the world."

The *Globe*'s Eric Reguly, as is often the case, has a much better understanding of what has actually been happening:

> In civilized countries, there is a sense of tax balance. Corporations and individuals pay taxes. The split is never 50–50, but the direction Canada is going is already totally out of whack. In 1961, corporate tax as a percentage of personal tax was 63 percent. Last year it had fallen to 32 percent. In other words, the relative tax burden on the individual has doubled.[6]

Newly revised figures now show that federal and provincial corporate tax as a percentage of personal income tax was actually down to only 29 percent in both 2005 and 2006, quite a drop from 63 percent. Put another way, in the huge profit year of 2006, corporate income taxes were only 8.8 percent of total federal and provincial government revenue.

Of course, when you consider corporate taxes, it's wise also to consider a topic that, remarkably, you seldom read or hear about: the already very high and growing levels of corporate concentration in Canada. And why don't we read or hear more about them? Read the chapter on the media in this book.

And what about the changes the Stephen Harper government was planning for its 2007 mini-budget? Reguly writes, "If anything, the pendulum is swinging even further in corporations' favour. They will pay relatively less, you will pay relatively more. If that doesn't sound fair, it's because it isn't."

In the same article, Reguly raises another topic almost entirely invisible in most of our media: tax losses from foreign takeovers. "Generally speaking, a foreign owner loads its new Canadian subsidiary with debt. Since interest payments are tax deductible, tax-bills will plunge."

And you can bet that the interest rates the foreign buyers charge their new subsidiaries will be sky-high. Reguly gives an example of a high-margin cigarette manufacturer in Canada with sales of $404-million that manages to pay taxes of only $10-million after paying the takeover firm $106-million in interest payments.[7] But have no fear, a *Globe and Mail* editorial tells us that cutting corporate taxes is "a superb way to attract foreign direct investment and give more money to Canadian companies for needed equipment and machines."[8]

Finally, guess just who it is that has to make up the lost tax revenue. You might want to look in the mirror and think about this every April.

Let's go back and do a quick summary of the rate of reduced taxes on net income paid each year by corporations.

2000	35.4%
2001	28.6%
2002	26.4%
2003	27.0%
2004	25.8%
2005	25.6%

Question: Do you think it's fair that in 2005, on their massive profits of $157.55-billion, corporations paid direct taxes at the rate of only 25.6 percent? Compare that to the tax rate you pay.

But wait, thanks to Thomas d'Aquino, Jack Mintz, and their friends in Ottawa, by 2012 federal corporate taxes will plunge to only 15 percent, the lowest of all major industrialized economies, and combined federal and provincial taxes will be far below the rates for the United States and Japan, for example. The Conservatives' 2007 mini-budget claims that their corporate tax changes "will increase Canada's statutory income tax advantage over the U.S. to 8.8 percentage points."

In 2006, the C.D. Howe Institute, true to form, said, "The pace of tax reform has been too slow," even though the corporate tax burden was by then already at its lowest level in almost 20 years, to a point where combined federal and provincial tax rates were already 3 percent lower than U.S. rates, and scheduled to be 5.8 percent lower in manufacturing and

3.3 percent lower for new investment by 2011. After the 2007 mini-budget, Tom d'Aquino said that the new lowered corporate tax rates gave Canada "an advantage of more than 12 percentage points over the United States."[9]

When comparative international figures have been published every year, the suggestion that corporate taxes in Canada were far too high have proved laughable. PricewaterhouseCoopers's comparison of total corporate tax rates back in 2005 showed that, in a list of 22 OECD countries, Canada came in way down in 16th place in the total tax rate as a percentage of profits. That put us below the United States, and below Italy, Belgium, France, Greece, Spain, Germany, Sweden, Austria, Japan, Australia, the Netherlands, Finland, Portugal, and Norway as well.

In their 2006 economic survey of Canada, the OECD pointed out that back in 2003, total corporate taxes paid on goods and services as a percentage of GDP in Canada were already almost 3 percent below the OECD average. In the list of 30 OECD countries, Canada was down in 27th place, far below the OECD average.

And by the way, the next time you read about the Fraser Institute's so-called "tax freedom day" – the day in the year when you have earned enough to pay all your annual taxes and the rest of your earnings no longer go to the government – you might let them know that they lose all credibility by intentionally basing their calculations on cash income instead of total income, which would make a large two-month difference in their calendar date.

The right-wing anti-taxers frequently go out of their way to pretend that total government revenue as a percentage of GDP is synonymous with tax as a percentage of GDP. It's nothing of the sort. Total government revenue includes many things not related to tax, things such as rental income, transfers from government enterprises, fines, fees, sales revenue, gambling revenue ($13.3-billion in 2006), etc., etc.

However, even if we are conned into using this very misleading and frequently used comparison, of the 30 OECD countries Canada was still down in 17th place, at 40.3 percent of GDP in 2006, well behind a group of countries in the range of 45 percent to 60 percent.[10]

And shouldn't we ask why the Fraser Institute and Canada's business press have failed to publicize the "tax freedom day" for corporations, which York University tax expert Neil Brooks and author Linda McQuaig point out would come every year in January!

One other point. You might also want to ask the Fraser Institute how they manage to get away with giving their wealthy right-wing corporate donors tax receipts. They manage to be a "charity," because they somehow claim that they are not engaged in political activities "to influence law, policy and public opinion" by way of conferences, speeches, lectures, publications, or published or broadcast statements. How the tax department allows this "charity" to get away with such blatant deception is totally beyond comprehension. Of course it's you, dear reader, who ends up involuntarily paying for the Fraser Institute's propaganda.

Jim Stanford suggests that "one option is to simply admit the corporate tax cuts are a failed experiment in trickle down economics." He puts corporate taxes into perspective:

> Federal corporate income taxes were cut by 7 percentage points between 2001 and 2005, saving companies at least $10-billion a year. They aren't spending the money on investment. So, let's channel it into needed capital spending . . . projects such as hospitals, colleges and universities, transportation and communications, infrastructure, etc. Everyone admits we need more of these long-term investments.

Or, why not use at least half of the increased tax revenue that we should be getting from corporations to reduce the tax load on individuals? There's a radical thought. Economist Don Drummond of TD Bank Financial Group calls the 2007 mini-budget's personal tax relief "paltry . . . it's just a derisory amount."[11] And in 2012-2013, the share of personal taxes as a percentage of GDP will go up even further, while corporate taxes to GDP will continue to fall. So individuals will pay four times as much tax as a percentage of GDP as corporations.

Oh, yes, for yet one more reason why the Senate of Canada should be abolished, simply consider their 2005 call for even more across-the-board tax cuts for corporations.

Before we turn to personal tax in Canada, let's again look at the supposed need to cut corporate taxes in relation to productivity and competitiveness. Of course our right-wing "think tanks" keep telling us that high taxes hurt our competitiveness. Interesting. The World Economic Forum and the World Bank both rank the Nordic countries at or near the top of their competitiveness lists. Yet Sweden, Denmark, Norway, Finland, and Iceland have high tax-to-GDP ratios. And productivity expert Andrew Sharpe has shown that eight "high tax" Western European countries have higher output-per-hour levels than the United States, while OECD forecasts for the largest current account surpluses invariably include "high tax" countries such as Norway, Sweden, Finland, the Netherlands, and Germany.

The 2005 analysis of the World Economic Forum is instructive. Four high-tax Nordic countries were ranked in the top 10 in competitiveness. These countries also have generous social services and very low rates of child poverty and are invariably ranked very high in quality-of-life surveys. In the same World Economic Forum survey, Canada ranked 15th in competitiveness.

If, as we have seen, corporate taxes in Canada have been steadily coming down, how do you explain the continuing poor business competitiveness and productivity ratings? Andrew Sharpe writes: "Growth in the machinery and equipment capital stock and capital-labour ratio has fallen off in recent years, reflecting slower investment growth. This has reduced the rate at which new productivity enhancing technologies are put into operation."[12] If this doesn't yet again make a mockery of big-business promises and performances, and the reasoning behind the need for even more corporate tax cuts, I don't know whatever would or could.

I recommend the paper *The Social Benefits and Economic Costs of Taxation: A Comparison of High- and Low-Tax Countries* by Neil Brooks and Thaddeus Hwong, published by the Canadian Centre for Policy Alternatives in December 2006. Brooks and Hwong's paper starts this way:

"I believe all taxes are bad." Stephen Harper made this remark during the federal election last year. . . .

Taxes are the price citizens of a country pay for the goods and services they collectively provide for themselves and for each other. So it is difficult to know exactly what Harper meant when he said he believes all taxes are bad. Was he saying that all action taken collectively by citizens through democratically elected institutions are bad?

Tax levels in Canada have always been substantially below those in most other industrialized countries, and they have been significantly reduced over the past few years, yet the crusade against them continues unabated. In the average European country, tax levels were almost 5 percentage points of GDP higher than those in Canada.

This document shows that the so-called "high tax" countries score better in 42 out of 50 social and economic measures than the low-tax countries, with lower poverty, better income distribution, much better worker economic security, more leisure time, less drug use, and a higher GDP per capita. People in these countries say they are more satisfied with their lives, they have a higher ranking in post-secondary achievement and innovation, and pension levels are higher. In all the higher-tax Nordic countries there is greater gender equality, lower rates of infant mortality, longer life expectancy, and homicide rates are lower.

By way of contrast, say Brooks and Hwong,

Americans bear incredibly severe social costs for living in one of the lowest-taxed countries in the world. For a strikingly large number of social indicators, the United States ranks not only near the bottom of a list of 19 industrialized countries, but it ranks as the most dysfunctional country by a considerable margin.

And back to the subject of productivity:

On average the "high-tax" Nordic country workers produce goods and services valued at $44.1 an hour, while Anglo-American workers only produce goods and services valued at $38.2 an hour.

On the subject of competitiveness,

Every fall, the World Economic Forum, a business-dominated, Geneva-based, private organization, releases its Global Competitiveness Report. The report contains a comprehensive index that measures the competitiveness of countries based upon around 150 variables, including each country's macroeconomic performance, the quality of its public institutions, and the level of its technological readiness. On its index of growth competitiveness, the high-tax Nordic countries are significantly more competitive than the low-tax Anglo-American countries (an average score of 5.66 versus 5.35). . . . The World Economic Forum concluded that "There is no evidence that relatively high tax rates are preventing these countries [the Nordic countries] from competing effectively in world markets, or from delivering to their respective populations some of the highest standards of living in the world."

The low-tax United States ranks as the sixth most competitive economy in the world, but the high-tax Finland was the second most competitive country in the world. In addition to Finland, two other Nordic countries also rank in the top five most competitive countries in the world, with Sweden as third and Denmark as fourth.

What should we have done instead of cutting the GST? After the 2007 mini-budget, James Travers of the *Toronto Star* put it this way: "Saving each of us pennies on coffee will cost us at least $34 billion over 5 years, money more profitably spent on health care, education, cities, poverty and the environment."[13] Or how about a proper child-care program, a national pharmacare program, some decent social housing? Forget it. Finance Minister Jim Flaherty has already said there will be no new spending programs of consequence.

Is cutting the GST good economics? The *Globe and Mail*'s Jeffrey Simpson says, "To present an economic package that offers $3 in consumption tax savings for $1 in personal income tax savings is an economic travesty, a resolute defiance of international practice, a wilful disregard of informed domestic advice, an economic nonsense and a political bet on economic illiteracy."[14]

As might have been forecast, the C.D. Howe Institute's former leader Jack Mintz has already been calling for even greater corporate tax cuts.[15] But instead, after what you've read in this chapter, and will read in the next about excessive rates of personal taxes in Canada, you might ask, Why not cut personal taxes? The reason . . . wait for this: according to Mintz, personal tax rate reductions are fiscally expensive.

PERSONAL TAXES

Between 1961/1962 and 1966/1967, the total of personal income tax in Canada was never greater than 5 percent of GDP. From 1977/1978 to 1985/1986, it was never higher than 6.9 percent, and most years it was between 6 and 6.3 percent. Then, beginning with the first year of the FTA, the personal tax share jumped to 7.7 percent of GDP and has been close to or over 8 percent in most of the years since. In another comparison, whereas in 1962/1963 corporations paid over 20 percent of total federal tax revenue, by 2005/2006 this was down to only 14.3 percent.

What about total federal government income tax revenues, personal and corporate, as a percentage of GDP? Again, contrary to conventional wisdom, at 10.2 percent of GDP for 2006/2007, they are below the average level for the years 1995/1996 to 2004/2005.[1]

A somewhat different perspective comes from looking at the share that personal income tax constitutes of total federal government revenues. In 1965/1966, it was as low as 28.9 percent. In 2005/2006, it was all the way up to 46.7 percent. By 2011, it is scheduled to increase to over 53 percent. In other words, once again we can see that there has been a major shift in taxes from corporate sector tax to personal income tax.

How does personal tax in Canada compare with personal taxes in other countries? Only seven OECD countries have higher personal taxes as a percentage of GDP: Australia, New Zealand, Belgium, Denmark, Finland, Iceland, and Sweden. But 22 countries have lower total taxes on personal income. Canada's rate is higher than the OECD and the EU averages and is the highest in all the G7 nations.

If we look at taxes on personal income as a share of total taxation, Canada has a higher percentage than all but three other OECD countries: Australia, New Zealand, and Denmark.

Before the most recent GST cuts, taxes on goods and services in Canada as a percentage of GDP in 2004 were 8.7 percent, while the OECD average was 11.4 percent. In these taxes, Canada is far down in 27th place.[2]

In 2004, per-capita tax revenue in Canada in U.S. dollars came to $10,552, compared to $10,147 in the United States, a much lower difference than what is commonly trumpeted by many of our right-wing, continentalist commentators and many of our politicians. Meanwhile, please note that the OECD average was $11,229 (U.S.), and the EU average $12,008 (U.S.).

In 1990, Canadian tax-filers, on average, paid $12.25 of federal tax for every $100 of income. By 2002, that had declined to $11.18.[3] Despite what *Maclean's* has told us, the top federal income tax rate in Canada has been 29 percent. The highest income earners pay an overall *effective* tax rate of 31.8 percent (2004), that is the tax *actually paid*, while "non-high income" filers paid roughly 12 percent,[4] and the overall effective personal rate was 16.3 percent. Again, this is quite different from what we often read in the press.

Let's compare personal taxes of workers in the manufacturing sector. Production workers in Canada, on average, are taxed below the OECD average and far below the EU15 average. As a percentage of labour costs, personal income taxes in Canada plus any employee and employer social security contributions worked out to just over 32 percent in 2004. The OECD average was 36.5 percent, and the EU15 average was 40.8 percent.[5]

Of the 30 OECD countries, Canada was in 19th place in these taxes. Belgium was highest on the list at 54.2 percent, followed by Germany at

50.7 percent and Sweden at 48 percent. Mexico was at the bottom of the pack at only 15.4 percent, much as might be expected, given the low wages and relatively poor social benefits in that country.[6]

Let's zero in on after-tax disposable income. Once again, contrary to the image presented so often by many of our politicians and much of our media, the disposable income of a single Canadian production worker as a percentage of gross pay (75.6 percent) is almost identical to the disposable income of an American production worker (75.7 percent).

This, of course, is interesting to consider when you look at how much higher Canadian social benefits are almost right across the board, but in particular when you compare personal and family health-care costs.

For a married couple with two children, the disposable income in Canada, at 85.7 percent, is only marginally below the U.S. rate of 88.5 percent. However, health-care premiums for the U.S. family cost many thousands of dollars and have been increasing rapidly every year, and at an accelerated rate in the past few years. Employer health insurance in the United States has been unravelling, with annual costs for a family of four now some $12,000 (see Appendix One). A key problem remains the extremely inefficient and costly fragmentation of the U.S. system, with administration costs of about 25 cents on every dollar. This compares with 2 cents for the Canadian single-pay system. Of the over 47 million Americans with no health insurance, almost half are full-time workers. Last year, there were almost 850,000 personal bankruptcies in the United States due to the inability to pay medical bills.

Let's do a quick further comparison with taxes in the United States. Despite the fact that Canada's total tax load leaves us down in 21st place compared to other OECD countries, we continue to get a steady cry from the same old right-wing gang that taxes in Canada are too high compared to those in the United States. Our taxes are, in fact, higher than those in only nine other OECD countries, but as we've seen, Canada's corporate tax rate has already fallen well below the U.S. rate, and our manufacturing tax rate on capital is 33.1 percent, compared to the U.S. rate of 34.8 percent. On services, we're down to 39.6 percent, compared to the U.S. rate of 40.2 percent. By 2005, the effective Canadian corporate income tax rate on manufacturing pre-tax profits was down to 32 percent, compared to the U.S. rate of just over 37 percent.

By the way, the same big-business Canadian Council of Chief Executives (CCCE) corporations constantly bellyaching for even more corporate tax cuts – while at the same time recording all-time-record after-tax profits – have seen their marginal effective tax rate on business investment fall from 44.6 percent in 2000 to 37.7 percent in 2005, with further reductions scheduled for 2010 to about the same level as the U.S. rate. In a July 2007 letter to the *Globe and Mail*, federal Finance Minister Jim Flaherty said, "We are moving Canada's overall tax rate on new investment from third-highest to third-lowest in the G7 by 2011."

The Canadian Labour Congress, in a 2007 presentation to a House of Commons finance committee, made the following important point regarding taxes on new investment: "Marginal tax rates are not the appropriate measure of international competitiveness. An investor deciding where to locate a facility is concerned about the investment's total tax liability, not the tax on the last dollar invested."

For our corporate continentalists and compradors who want us to copy everything that happens in the United States, it's worth asking if they would be pleased if we continued to drop our total taxes down below U.S. levels, and, by the way, how do they like the unprecedented U.S. government deficits and their rapidly ballooning debt? Perhaps ask them how they would pay for that debt, which is continuing to grow every year, with more big budget deficits guaranteed and forecasts by the Congressional Budget Office for hundreds of billions of dollars more debt a year over the next decade, with an accumulated debt headed for $10-trillion by 2009.

Whatever else you may want to say about them, the flacks for corporate Canada have been enormously successful in their campaign for lower taxes. For effrontery, you can't beat this, from Thomas d'Aquino of the CCCE: "Lower corporate taxes pay for all Canadians."

In the Conservatives' 2006 budget, corporate manufacturing statutory tax rates were to be cut to 5.1 percent below U.S. rates by 2010, and Canada's marginal effective tax rate will be lower than the U.S. rate.

Quickly, on to some other tax topics. Let's start with one that costs you personally, as a taxpayer, a bundle every year – tax havens.

In March of 2005, Statistics Canada revealed that between 1990 and 2003, Canadian assets in tax havens increased from $11-billion to $88-billion. The largest growth was in Barbados, Bermuda, the Cayman Islands, the Bahamas, and Ireland. Most of the assets were in the financial sector.[7] A 2004 University of Quebec study said that since 1991 our major banks have used tax havens to avoid paying $10-billion in taxes. One 2005 estimate by the Tax Justice Network put the tax revenue lost to tax havens by governments around the world at $255-billion (U.S.) annually.

Have you ever wondered why Canadian direct investment abroad is so high in the figures trotted out in an attempt to balance concerns about the buyout of Canadian industries by foreign companies? Here's something you probably didn't think of. A large and increasing share of that "investment" is tax avoidance. Money is being sent to tax havens at a rate now double what it was a decade ago. Canadian direct investment abroad in "offshore financial centres" increased eightfold between 1990 and 2003 to $88-billion, or about a fifth of all Canadian direct investment abroad. As you probably know, Paul Martin's Canada Steamship Lines paid some of their taxes in Barbados at a 2.5 percent tax rate. And, of course, every dollar Canadian governments lose in tax havens is a dollar that comes out of the pockets of readers of this book.

By the way, a favourite trick of our corporate and "think tank" anti-taxers is to include a long list of tax havens in their overall combined list of country tax comparisons, giving, for example, the same tax weight to, say, the Cayman Islands as the United States, producing highly misleading results.

There are those, including myself, who have long suggested increasing the basic income tax exemption so fewer low-income Canadians have to pay tax. But as Roger Martin, dean of the Rotman School of Management at the University of Toronto, has aptly pointed out, increasing the basic personal income tax exemption across the board is poor public policy, because as much as 67 percent of the tax benefit will go to people who are not low-income earners. Says Martin, "It's a poor way to help the target group."

Also poor public policy are the 2006 Conservative tax changes which saw the highest-income families, the top 5 percent, get 28 percent of the tax savings, while the lowest half of Canadian families saved only 20 percent.

While recent federal tax changes are nowhere near as lopsidedly in favour of the already well-to-do as they are in the United States, there's no question that they mostly benefit the well-off while ignoring good public policy such as increases to the Child Tax Credit for low-income families, or an earned-income supplement for the working poor.

Meanwhile, incredibly, one-third of Canadian families with incomes of $100,000 or more are receiving a GST credit.[8] In fact, only a quarter of the families receiving the GST credit in 2003 were classified as low-income.[9] How's that for great public policy?

I can't leave the topic of taxes without a brief discussion of Ireland. If you pay any attention to the anti-taxers, Ireland is the shining new example of what low taxes can do for a country. Neil Brooks and Thaddeus Hwong in their CCPA study say,

> There is little reason to suppose that tax cuts had much to do with the Irish economic miracle. Ireland reaped the advantages of huge European Union subsidies, particularly in the late 1970s and in the 1980s (reaching 6 percent of GDP), and even in the early 1990s. Ireland invested those subsidies in infrastructure, including free higher education. It had an English-speaking, well-educated, under-utilized labour force. . . .
>
> It seems reasonably clear that the Irish miracle is due to a unique set of circumstances that cannot be duplicated in other countries simply by trying to imitate its beggar-thy-neighbour corporate tax rate strategy.

What I have found almost amusing and yet disgusting at the same time are the numerous corporate, individual, and media complaints about how the well-to-do have been paying more than their share of the tax burden. But what the press reports often somehow fail to mention is that the same people pay a higher share of taxes because they take in much more than half the income and have much more than half the wealth, as you have seen in the preceding chapters.

I have been writing and speaking about the major problem caused by excessive foreign ownership and tax revenue lost in transfer pricing in Canada for years, but alas there is no room to do this important topic justice here. In 2004, there was finally a major court decision that shed further light on the problem. Mr. Justice Peter Cumming of the Ontario Supreme Court ruled that the Ford Motor Company had been cheating Canadian shareholders for years by unfairly transferring profits out of the country, disregarding the interests of Canadian investors while the U.S. parent company manipulated prices on vehicles and parts sold to and purchased from its Canadian subsidiary to avoid Canadian profits and taxes. My own educated guess is that transfer pricing has cost Canadians close to $20-billion a year in lost tax revenue. (For an excellent article on transfer pricing see the March 2007 issue of the CCPA's *Monitor*, written by a former vice-president of Pfizer, Peter Rost, who knows something about the topic.) Of course, while corporate taxes are now lower in Canada, transfer pricing for tax reasons should be less of a problem, but foreign firms will undoubtedly continue to charge their Canadian affiliates unfairly for goods and services, if only to avaoid sharing profits with their Canadian shareholders.

As for the great baloney from the Fraser Institute, Conrad Black, and the *National Post* on the supposed brain drain from Canada to the United States, late in 2006 Statistics Canada supplied even more evidence showing that the number of men and women leaving Canada was exceedingly low, about 0.1 percent in any year, and that the numbers have been dropping substantially in recent years. In the words of Queen's University's respected economist Ross Finnie, "What people thought was rising inexorably has come down substantially – even dramatically."

My only regret is that the once-departed Conrad Black has indicated that he would eventually like to regain his citizenship, though I personally would not regard his permanent departure as any kind of a lamentable brain drain.

The Harper government has announced plans to reduce the GST from what was once 7 percent to 5 percent. While this may be good politics, it is not very good economics if we look at the available alternatives. For example, as we have seen, personal income taxes in Canada are higher than the OECD average and higher than the EU15 average. However, if we look at federal and provincial taxes on goods and services as a percentage of GDP, of all the 30 OECD nations, 26 have a higher tax. Canada, at only 8.8 percent in 2003, was far below the OECD average of 11.5 percent, and even further below the EU average of 12.2 percent.[10]

So, since we were already way down in 27th place in taxes on goods and services, and since our personal income taxes have long been so high, why cut the GST to even lower levels? True, someone buying expensive jewellery or an expensive home or new Bentley will save a bundle, but a 1 or 2 percent saving on even inexpensive household items represents only pennies. Yes, pennies to a poor person are important, but 2 percent of the cost of a million-dollar house could pay for a big pile of groceries for many poor families.

When the GST dropped from 7 to 6 percent, the cost to the national treasury was in the billions of dollars a year. Imagine what that sum could do for a child-care program or for housing for the homeless, or to help feed hungry children, or to provide more scholarships for low-income students, or to help implement the abandoned Kelowna Accord.

The anti-taxers, in the words of Carol Goar of the *Toronto Star,* "refuse to believe that the public services they get – health care, good schools, reliable pensions, clean water, livable cities – are worth the taxes they pay." What organizations like the C.D. Howe Institute and the Fraser Institute (both essentially funded by the very same big-business corporations) have been able to do is disconnect the benefits we receive from the taxes we pay.

Linda McQuaig puts it well:

Clearly, we can't have the kind of public services Canadians say they want without paying for them. Just as there's no free lunch, there's no free health care – or free garbage pickup. But this fundamental fact is omitted from the fantasy world presented to us by the right-wing tax-cutters.

Meanwhile, the people of Europe and Scandinavia pay a lot more tax, but they get a lot more services – like national day care, extensive parental leaves, comprehensive home care and drug programs, free dental care for children, free university tuition – things that would be dismissed here as pie-in-the-sky dreaming.[11]

Aside from all the benefits such as highways, bridges, parks, schools, colleges and universities, police, utilities, employment insurance, job training, garbage collection, programs for seniors, child benefits, and a long list of other public services, Statistics Canada tells us that in 2004 average taxpayers received $15.68 in government transfers for every $100 in employment income.[12] Put another way, median family market income in 2004 was $55,800 and median income taxes were $8,600, but government transfers left net income of $51,200, or close to 92 percent of market income.[13]

For many of our most strident anti-taxers, civilization is a gated, privately patrolled community in Florida, California, Arizona, or Hawaii. But the vast majority of Canadians have something entirely different in mind when they consider the quality of life in a desirable society.

Lastly, on the subject of corporate taxes as opposed to personal taxes, the next time you read yet another one of big business's never-ending pleas for even more reductions in their share of taxes, I hope you will consider the following important words from the OECD *Observer* of May 2007:

Business keeps finding ever more creative ways of getting around paying any taxes at all. . . . If more attention isn't given to this issue, we could soon face a global tax crisis.

While corporate tax rates are falling, corporate profits are booming and wages are stagnating. After-tax profits in the U.S. are, as a proportion of GDP, at their highest in 75 years and in Japan and the Euro area they are close to 25-year highs. Meanwhile, wages in the U.S. are at their lowest level since 1966.

Relying on income and spending of wage earners to fund ever larger parts of public financing will either hollow out government budgets or lower worker incomes. For the sake of both equity and efficiency, business tax cannot be allowed to go on falling.

The upshot of inaction will be a loss of revenue for governments and a downward spiral of economic activity.

HOW COMPETITIVE IS CANADA?

"The lower our taxes get, the less competitive we have become."

I n 2001, the World Economic Forum (WEF) ranked Canada third in the world in competitiveness. By 2007, after a series of major corporate tax cuts, we were down in 13th place. But economist Jim Stanford put it in proper perspective:

> In business lexicon, "competitiveness" is typically understood as synonymous with "low taxes." Ironically the lower our taxes [in Canada] get, the less competitive we have become.
>
> Nine of the 15 countries ahead of us on the WEF list collect higher taxes than Canada. Some of these countries rake in 50 percent or more of their GDP.
>
> Back in 1999, when we ranked fifth on the WEF scorecard, Canada's taxes were slightly higher than the OECD average. Today they are substantially lower. Canada's taxes have fallen faster since 1999 than any of the 15 countries ahead of us.[1]

The WEF has an explanation: "Countries that, like the Nordics, are investing heavily in education are likely to see rising levels of income per capita, growing success in reducing poverty and an increasing ability to establish a presence in the global economy."

Stanford points out that education budgets in Canada are lower today as a share of GDP than in 1999, and more than 20 percent below 1993 levels. Now, how do we pay for education? Oh, yeah: taxes.

Stanford continues:

> The utter lack of correlation between taxes and competitiveness did not stop Canadian business commentators from ascribing our weak performance to (what else?) high taxes and demanding still more cuts.
>
> The failure of Canadian competitiveness is primarily due to the failures of our businesses.

Toronto Star business and economics columnist David Crane quotes renowned Harvard University competitiveness expert Michael Porter to the effect that competitiveness "is a measure of how well a country uses the skills of its people, its investments in technology and infrastructure, its natural resources and its knowledge to efficiently produce goods and services others want to buy in open markets."

The WEF report says the top-ranked country, Switzerland, "reflects a combination of a world-class capacity for innovation and the presence of a highly sophisticated business culture." As Crane points out, by way of contrast, "One reason for foreign takeovers of Canadian companies is the lack of management skills by Canadian executives."

The 2007 WEF report ranks Canada down in 20th place in business sophistication.

True. But there are many other very important reasons for so much foreign ownership and control, which I will discuss in the chapter on foreign investment in Canada.

Worth considering before we go on to the chapter on education is the fact that high-tax Finland has placed number one or number two on the WEF competitiveness list for five of the last six years, and Finland leads everyone in early-childhood development and ranks very high in numeracy and literacy skills in all its socio-economic groups.

That Canada is so far down the competitive list when our labour costs, construction costs, employee wages and benefits, land and industrial-space costs, utilities and infrastructures, rental costs, our abundance of resources,[2] and so on, are so very competitive would be difficult to explain – unless you've read the preceding chapters on the failure of our big-business leaders to invest and innovate. In 2007, average business investment per worker in Canada was far below U.S., G7, and OECD levels.[3]

PART FIVE

EDUCATION IN CANADA

A GREAT SUCCESS, AND REMARKABLE FAILURES

I f you were shocked by any of the comparative international figures you've read so far in this book, wait until the last paragraphs in this chapter, when you see how we compare in education funding.

The OECD says, "The share of the population that has attained qualifications at the tertiary level is a key indicator of how well countries are placed to profit from technological and scientific progress. Tertiary programs are designed to provide qualifications for entry to advanced research programs and professions and/or are intended to lead directly to employment."

So, if we measure the number of people aged 25 to 64 years old who have attained a tertiary, in other words post-secondary, level of education, how does Canada do? The answer is very well indeed. In fact, Canada is in first place, ahead of all the other OECD countries, a big improvement over our fourth-place ranking in 2000. At 54 percent of the population in 2005, Canada was well ahead of the United States, at 39 percent, Japan at 37 percent, and Sweden at 34.5 percent, and well ahead of the OECD average of only 32 percent.[1] We have steadily increased our percentage of students with post-secondary qualifications, from just under 30 percent in 1991 and 34 percent in 1998.

If one looks only at the 25 to 34 age group, Canada is still in top spot, but the United States slips to eighth. In this group Canada, at just over 53 percent, is far ahead of the OECD average of 31 percent. Only Japan comes close to Canada in this age group. In a less definitive study, Statistics Canada said that in 2005, 72 percent of Canadians aged 25 to 34 had "some type" of post-secondary education, compared to only 54 percent in 1980.

While Canada has the highest rate of post-secondary education among all 30 OECD members, it's important to note that much of our high ranking comes from enrolment in colleges. Looking only at graduation from universities, we're in sixth place, tied with Australia and Korea at 22 percent, while the United States is first, at 30 percent. Canada is ahead of the OECD mean of only 15.5 percent, but also well behind Norway at 28.4 percent.

This said, on a per-capita (and troubling) basis, we produce only about half as many MAs and about one quarter as many PhD graduates as the United States. In 2006, only 5.8 percent had a degree above a bachelor level. The Institute for Competitiveness and Prosperity at the University of Toronto has estimated that the current gap costs Canada some $30-billion annually.[2]

In 2004/2005, enrolment in Canadian universities surpassed the one million mark for the first time. It was the seventh consecutive year that a new enrolment record was set. (University enrolment had previously been on the decline in the mid-1990s.)[3] In 2004/2005, there were over 413,000 university students enrolled in Ontario, over 263,000 in Quebec, over 88,000 in Alberta, and 87,000 in B.C.

Let's now see how effective our other school systems are. The OECD has a Program for International Student Assessment to measure student knowledge and skills in mathematics, science, reading, and cross-curricular competencies at age 15. The survey involved 3.5 hours of testing time in mathematics, and one hour each for reading, science, and problem-solving.

As in the past, Canada once again comes out very well. In reading skills, we're fourth overall, well above the OECD average, and also above the United States. In science, we're third, once again above both the OECD and U.S. averages. In mathematics, we're seventh, while the U.S. is far down in 24th place. In another slightly different international mathematical test, Canadian students ranked third, behind only Finland and Korea.

South Korea tops the reading scale, followed by Finland and Hong Kong. In science, Finland and Hong Kong lead the way. In math, Taipei, Finland, and Hong Kong are tops.[4]

Certainly, it's interesting to note that in Canada school dropout rates have declined significantly, from about 17 percent during the 1990/1991 school year down to 9 percent by 2005/2006. And many dropouts (27 percent by one study) return later to complete high school. Of those who do return and graduate, about half go on to post-secondary studies. This said, Canada's high school dropout rate is still higher than in many OECD countries and twice the rate for Norway.

Let's look at the percentage of foreign students in our universities and colleges, and see how they compare with other countries. Seven OECD countries have a higher percentage of foreign students as a percentage of all post-secondary enrolment. New Zealand is by far the highest, at 28.3 percent, followed by Australia at 19.9 percent and Switzerland at 18.2 percent. Canada, at 7.4 percent, is slightly ahead of the 7.3 percent OECD average, and even further ahead of the 3.4 percent for the United States,[5] but in 2005 we attracted only 2.8 percent of cross-border tertiary students.

In 2004/2005, a record 75,200 students from other countries were enrolled in Canadian universities, up 7.3 percent from the previous year. Half the foreign students were from Asia, and over 46 percent of these were from China.[6] Some 17 percent came from Europe, and 18.5 percent from the Americas, including the United States. This said, where Canada was once in the top five as a destination for foreign students, we're now 14th.

Before we go on to look at tuition costs and government funding, let's consider a growing and serious problem at our universities. Queen's University's Kim Richard Nossal, president of the Canadian Political Science Association, points out that from 1976/1977 to 2003/2004, full-time enrolment in Canadian universities almost doubled with only a tiny increase in faculty. In political science, there were almost 9,000 more students, but only 27 more professors! Another political science problem is that in several important Canadian universities, such as the University of British Columbia, the University of Toronto, and McGill University, American-trained academics are in the majority: 59 percent at U.B.C., 69 percent at U. of T., and 76 percent at McGill.[7]

In Ontario, which has the worst student-faculty ratio in the country (24 to 1 compared to the national average of 18 to 1), universities are threatening to turn away even more aspiring entrants or have even greater levels of classroom crowding. For fall 2003, the University of Ottawa received 51,000 admission requests (many of whom, of course, also applied elsewhere), but accepted only 6,000 new students. For fall 2007, University of Calgary required entrance grades were at an all-time high of 85 percent.

More and more qualified students have found it difficult to get into a university of their choice in Canada, while at the same time tuition fees have risen so dramatically that these fees have become a major problem. In 1990/1991, undergraduate students paid an average of $1,469. By 2007/2008, they were paying an average of $4,524 (in Quebec, undergraduate fees were less than half the national average at only $2,025). There has been some suggestion that future tax credits will help alleviate the problem of high tuition fees, but as Ian Boyko of the Canadian Federation of Students said, these are "back-handed measures. They may help people recoup some of the costs, but don't improve access to post-secondary education."[8]

In May 2007, a report by the Canada Millennium Scholarship Foundation said that forgone tax revenue would be much more effective if the money was given directly to students in grants or loans, and warned that the system of future tax credits mostly benefits high-income families whose children would likely be going to university anyway.

The bottom line is clear to see. The full-time post-secondary enrolment rate for those with parental income over $80,000 is almost double the rate for those with parental income under $40,000. In another comparison, in the 1990s, about 40 percent of young Canadians from families with an income of $100,000 or more had a university degree. That compares with only about 19 percent from families with an income of $25,000 or less. By 2005, the percentage from well-to-do families had increased to 46 percent, while those from lower-income families inched ahead to 20 percent. In 2003, the most recent year for which statistics are available at this writing, only 31 percent of 19-year-olds from families in the bottom 25 percent of income distribution had attended university, whereas 50 percent of the same age group from the top half of income distribution had.[9]

It's interesting to note that in the 2006 edition of the OECD's excellent *Education at a Glance* (some "glance," at 449 pages), "Finland and Canada as well as five out of the six East and Southeast Asian countries . . . are among the countries in which social background has the smallest impact on student success. This suggests that these education systems succeed better in creating meritocracies that maximize the human potential of their countries more effectively" than countries such as Germany, France, Italy, and the United States. Once again, this said, almost a third of families polled in Canada say that financial difficulties are the main reason that family members do not pursue higher education.[10]

It's also interesting to note that for today's 25- to 64-year-olds, the average Canadian spent about 13.2 years in the education system. Norway is tops among OECD countries at almost 14 years, followed by Germany, Denmark, the United States, Luxembourg, and then Canada, followed by 24 other countries, where the length of time the average student is in the system goes all the way down to about 8.5 years in Mexico and Portugal.

It's worth underlining here that many EU countries have either no post-secondary tuition fees or only token fees. There are no tuition fees in the Czech Republic, Denmark, Finland, Iceland, Norway, the Slovak Republic, and Sweden. In the following countries, tuition fees average less than $1,000 (U.S.): Austria, Belgium, Hungary, Italy, Portugal, Spain, and Turkey.

By contrast, in 2006, the average university student debt on graduation in Canada was over $24,000. In Quebec, the province that had the lowest tuition fees and the most generous grants programs, the average debt was less than $13,000. Average undergraduate tuition fees for full-time university students in 2006/2007 were $4,347. For law students they were $7,221, for medicine $10,553, for dentistry $13,463.

Anyone who has spent time with university students across Canada knows the huge problem rising tuition has been causing in recent years. With such substantial tuition increases, ever-growing numbers in crowded classrooms, and more and more students forced to leave the country to enrol in university, it's time to examine funding for post-secondary education.

First, let's look at Ottawa's role. In real per-person dollars, federal government transfers to the provinces for post-secondary education, even with the Harper government's $800-million increase in the 2007 budget, are an estimated $1-billion less than they were 15 years ago.

In May 2007, the Certified General Accountants Association of Canada said that

> the quality of education has been experiencing a decline over the last two decades, largely as a result of poor investment levels on the part of provincial and federal governments.
>
> Federal cash transfers to the provinces for post-secondary education, for example, fell to 0.19 percent of gross national product in 2004–05 from 0.56 percent in 1983–84 . . . and while student growth increased nearly 50 percent between 1987 and 2003, faculty growth lagged at 7 percent.[11]

University of Alberta President Dr. Indira Samarasekara has pointed out that since 1992 Canadian universities have added 222,000 more students, but only 2,000 more faculty.

Meanwhile, public opinion polls show that 70 percent of Canadians believe Ottawa should increase funding for post-secondary education, two-thirds say university tuition fees have reached an unacceptable level, and 50 percent say that the federal government should supply a free university or college education to any qualified student who cannot afford it. In 1990/1991, student university fees contributed 12 percent to university incomes. Today, it's well over 21 percent. And one study estimates that by 2020 a four-year university education in Canada will cost $90,000.

Some other interesting numbers. Between 1980 and 2004, public university funding in the United States (in constant dollars) increased by 25 percent. In Canada, it fell by 20 percent. One study showed that public funding of post-secondary education in Canada on a real per-student basis declined by 30 percent between the 1980s and 2004. University of Toronto president, Dr. David Naylor, says that 20 years ago Canadian universities received $2,000 per student more from government than their U.S. peers, while today they receive $5,000 less.[12]

Here is one of the places where Canada does very poorly. In September 2006, an OECD report said that out of 14 industrialized countries, Canada was dead last in public spending on early childhood education as a percentage of GDP. At just a quarter of one percent, Canada's early-childhood spending was only one eighth that of Denmark, and far below the spending in Sweden, Norway, and Finland. Even the U.S. rate was double Canada's.

David Crane of the *Toronto Star* comments:

> The OECD review reports that among societies most concerned with their future competitiveness and about the social benefits of improving life chances for all children, there is "a growing consensus – based on research from a wide range of countries – that government on a cost-benefit analysis must invest and regulate early childhood education and care."[13]

Great, except governments in Canada, for the most part, have shown they aren't much interested. In contrast, in more than half of the OECD countries, 70 percent of children ages three to four are enrolled in either pre-primary or primary programs.[14]

Returning to the question of university tuitions, some people suggest that we needn't worry too much about rising fees because we have very generous programs for student loans, grants, and scholarships. We do, compared to some countries, but all of the following have more generous programs: Norway, New Zealand, Australia, Chile, the Netherlands, Sweden, Iceland, and the United Kingdom.

Others say there is little evidence that high tuition fees are an inhibiting factor for potential students from low-income families. One Statistics Canada study suggested that the gap between the attendance of affluent and poor young people was not so much a matter of money as it was reading scores at age 15, high school marks, and parental education. I have grave reservations about the logic here. If indeed it is correct, then surely a reading of the preceding chapters in this book on poverty indicates all the more reason why Canada should increase social spending to at least the EU averages. Young people growing up in poverty, with poor shelter, empty bellies, and inadequate clothing, are hardly going to be as well prepared for school tests as students from affluent families. The elitist conclusion of the study seems to somehow forget the inability of many poor children to properly compete on any kind of a level field. Of course there are exceptions, but they tend to be exactly that – exceptions. A 2006 public opinion poll found that more than three-quarters of Canadians believe that low-income students have less opportunity than high-income students to attend post-secondary institutions in this country.

Anyway, not to worry. In a February 2007 editorial, the *National Post* tells us, "If anything, tuition fees in Canada should be *increased*." Meanwhile, a spring 2007 poll showed that students surveyed as to why they were not pursuing post-secondary education cited financial issues as their number one reason.

A look at what happens in elite U.S. universities is interesting. A study released in the fall of 2006 showed that no fewer than three out of every five students at elite American universities gain entrance because of wealthy or alumni parents or because of "sporting prowess." Only 3 percent of the students in these universities came from the bottom income quintile, and only 10 percent came from the bottom *half* of the income scale.[15]

On another topic, sadly many students graduating from our schools and our post-secondary institutions have little knowledge of our own country. The excellent 2005 book *What Canadians Think,* by Darrell Bricker and John Wright of Ipsos Reid, showed that we're faced with a terrible lack of knowledge about Canada among our own citizens. Some examples:

- only 37 percent of Canadians could identify the first line of our national anthem;
- only 45 percent knew that Confederation was in 1867;
- only 47 percent knew about D-Day and Canadian participation;
- only 31 percent were able to name Dieppe as the French seaside town where almost 1,000 Canadians lost their lives in the infamous 1942 raids.

In November 2006, a Dominion Institute survey found that more than one in four Canadians thought U.S. General Douglas MacArthur was a Canadian, and only 31 percent identified Billy Bishop and Sir Arthur Currie as Canadians. In a list of questions including simple knowledge of Vimy Ridge and the poem "In Flanders Fields," 60 percent of those surveyed failed the test, and only a third could identify the four political parties represented in the House of Commons. In 2007, fewer than half of Canadians aged 18 to 24 could name Sir John A. Macdonald as our first prime minister, and three-quarters polled could not give the date of Confederation, while only 12 percent were able to name the Canadian prime minister who won the Nobel Peace Prize.

Good grief! It's been at least 30 years since I first raised the issue of poor Canadian content in our schools after doing a survey of over 3,000 students in their last year of high school in all 10 provinces. In the interval, many others have complained about the same problem. I can think of no other country that would so ignore its own history, its own heroes, its own culture, its own accomplishments to such an appalling extent. It's hard to imagine, but only four provinces require students to take a Canadian history course before they graduate. This said, I can't end this chapter without paying tribute to Avie Bennett, Peter Lougheed, James

Marsh, Charles Bronfman, and Red Wilson of the excellent Historica Foundation (www.histori.ca), which is doing so much to support programs and resources, in both French and English, to encourage Canadians to explore their own fascinating history and culture.

In conclusion, I don't know of any expert in post-secondary education who doesn't think that our universities are seriously underfunded. And new forecasts suggest we're going to have to plan for up to 170,000 additional new students in the next decade over and above the 815,000 full-time students enrolled in our universities in 2007.[16] At the three major universities in Toronto alone – York University, the University of Toronto, and Ryerson University – projections indicate a future need for between 40,000 and 75,000 new spaces, leading to talk of a whole new university.

At the beginning of this chapter I warned that you would be shocked by how Canada's spending on education compares to other countries. The latest edition of the United Nations *Human Development Report* reveals these truly appalling numbers for the period 2002-2005. In public expenditure on education as a percentage of GDP, Canada is way down in 57th place. In public expenditure on education as a percentage of all government spending, Canada is 90th.

90th!

Great numbers in a rapidly increasingly competitive world, especially when all signs indicate record numbers of students will be applying for post-secondary institutions in 2008.

CULTURE IN CANADA

"Bleed and starve Canada's cultural institutions until they croak."

For us to largely ignore our own history and identity when we live next door to the world's most aggressive culture-exporting country is a foolish mistake. Nevertheless, Canadian politicians seem determined not to provide adequate support for our own cultural community.

Jeffrey Simpson, writing in the *Globe and Mail,* put it well: "There cannot be another country in the world that makes it so difficult for domestic creators to find an audience."[1] For example, Air Canada is "a Canadian-owned airline that almost never showed a Canadian-made film." And, "Can anyone imagine a store window of Barnes and Noble in Chicago or Foyle's in London filled with imported trash from other countries?"

Margaret Atwood, writing in the *Globe and Mail,* laments the poor Canadian government support for culture, noting the apparent intention of the Harper neocons to "bleed and starve Canada's cultural institutions until they croak." This, despite the fact that we "now have an artist stimulated 'creative economy' that's worth – so they say – $40-billion a year. Why invest money in the arts? Because – simple answer – it's a great investment. A few dollars in means a lot of dollars out. Without the arts, the average Canadian citizen would be poorer and I don't mean just spiritually."[2]

Atwood has previously pointed out that there are more direct jobs in the cultural industries in Canada than in agriculture and mining combined. I think you can throw in forestry as well.

This will come as a great surprise to many Canadians. But every year, total spending on live performing arts in Canada exceeds spending on sports. The same is true for spending on books. In 2005, Canadians spent $1.2-billion to attend live performing arts presentations, more than the amount spent attending paid-admission sports events, and another half-billion was paid for admission to museums.[3] The same year, the book publishing industry in Canada had total revenues of $2.4-billion.

Yet when we measure total support for culture, it's under 1.5 percent of the total annual budgets of all three levels of government. Meanwhile, culture represents some 5.8 percent of GDP, and an enormous 94 percent of Canadians say we should be doing much more to promote our own culture.

It's interesting to look at household expenditures on recreation and culture (including spending on newspapers, sports equipment, gardening, movies, books, CDs, DVDs, toys, pets, etc., etc.). If we do this, as a percentage of GDP, Canada is in 11th place in the OECD. At the bottom of the list comes Mexico and, surprisingly, Ireland, both far behind all the other OECD countries. In 2005, Mexico spent only 1.9 percent of GDP per household on broadly defined culture and recreation, Ireland only 3.1 percent. At the top of the list, U.K. households spent 7.7 percent, Australia 6.9 percent, and the Czech Republic, Iceland, the United States, Austria, and New Zealand were all at or above 6 percent, while Canada was at 5.5 percent.[4]

Let's look at the sale of books, music, and newspapers in Canada. In total dollar value of book sales, Canada stands 10th in the world. Most of the countries with larger sales than Canada also have larger or much larger populations: the United States, Japan, Germany, the United Kingdom, Mexico, China, France, Italy, and Spain.

On a per-capita basis, measured in U.S. dollars per head, Canada is 16th in book sales, at only about $63 per annum. Norway is number one at $167, followed by Japan at $163 and Germany at $148. The other countries with per-capita annual books sales higher than Canada are Finland ($124), Belgium ($117), Switzerland ($112), the United States ($110), Singapore ($107), Sweden ($99), New Zealand ($92), the United Kingdom ($90), Denmark ($87), Ireland and Australia ($77), and Spain ($74). Of interest is the fact that Canada's per-capita book sales are ahead of such countries as Italy and France.

Now, if we look at total combined per-capita book, newspaper, and music sales in a list of 30 countries, Canada is down in 20th place. At the top of the list, in descending order, are Japan, Norway, Sweden, and Finland. Other countries with higher per-capita sales than Canada include Switzerland, Denmark, the United Kingdom, Austria, Germany, the Netherlands, New Zealand, and the United States.[5]

In CD and DVD sales, Canada is in sixth place in total dollar sales, but only in 17th place in per-capita sales. The United States is far ahead of every other country in the dollar volume of CD sales, more than double second-place Japan, but Norway is first in per-capita sales, while the United States slips to third place behind the United Kingdom.

Looking at the sale of daily newspapers measured in terms of copies per 1,000 population, Japan is in first place with 551, followed by Norway with 544 and Sweden with 481. Canada, way down in 20th place, is at 157, behind the United States at 188. Other countries with higher newspaper sales than Canada include Finland, Bulgaria, Switzerland, Denmark, the United Kingdom, Austria, Germany, the Netherlands, Singapore, Luxembourg, Hong Kong, Estonia, New Zealand, Slovenia, and the Czech Republic. Surprisingly, Canada is well ahead of Italy (138), Ireland (136), France (134), Australia (110), and Spain (105).

Meanwhile, some 85 percent of the space on Canadian newsstands is devoted to foreign periodicals, and yet our right-wing continentalist plutocracy has been enthusiastically lobbying to allow non-Canadians to take over our bookselling and book publishing industry. Foreign books already command almost 65 percent of the book market in Canada and two-thirds of the educational market, but Canadian publishers (with only 35 percent of sales) publish 85 percent of the Canadian authored trade books published in this country.

Canada imports more foreign book titles every year than any other country in the world, and more foreign magazines, almost all from the United States. In the fall of 2005, it was estimated that 650,000 different book titles were on sale in Canada during the previous 12 months. Just under 16,800 new titles were published in Canada during that time, and some 12,400 other titles were reprinted.

In 2005, Canada's trade deficit in cultural goods grew by 8.4 percent, to just under $1.67-billion, the largest increase in six years. The biggest deficit was with the United States, at $941.6-million. In 2006, the deficit grew again, to $1.8-billion, of which $1.2-billion was with the United States.

Our deficit in written material, especially periodicals and books, was by far the biggest contributor to the overall cultural-goods deficit. In 2006, imports of written materials accounted for about 73 percent of total cultural imports, while film and videos made up 7.3 percent. U.S. products accounted for about 78 percent of our cultural imports, and of these, books, newspapers, and periodicals represented some 76 percent of the total.[6]

A bit more about books. According to Statistics Canada, contrary to conventional wisdom, the purchase of books in Canada is the third highest cultural spending category, just behind newspapers at $1.22-billion and movie theatre tickets at $1.18-billion. Book purchases were, as mentioned earlier, well ahead of the dollars spent on live performing arts events and far ahead of the amount spent on attending sports events. All of this said, fewer than one in two Canadian households purchased even a single book in 2001.

In Canada, about 18 percent of English-speakers – described as "the book market" – buy on average more than six books in a three-month period, but on the other hand, 45 percent of English-speaking adults rarely buy books at all. In June 2005, Heritage Canada reported that, on average, Canadians spend 23 hours a week watching TV or listening to music, and only 4.6 hours reading. Mind you, close to 90 percent of Canadians said that they read books, more than half said that they read books every day, and 60 percent claimed to have read at least one book a month.[7] We must have very busy libraries.

Poor Canadian booksellers. Because their customers read reviews in U.S., British, and Canadian magazines and newspapers, they expect our bookstores to stock a full array of American and British books in addition to those published in Canada. One result is that Canada, year in year out, imports more English language book titles than any other country. In 2005, U.S. and British book publishers between them produced some 378,000 new titles.

In October 2006, a *Globe and Mail* reporter told readers that in the previous five years people had been devoting less time to reading as a leisure activity. But between 1993 and 2003, the most recent years for which Statistics Canada figures are available at this writing, we see that "Canadian [magazine] publishers are pumping out more periodicals than ever before and pulling in far greater revenues."[8] From 1998 to 2003, magazine industry revenues were up 22.5 percent, and profit margins increased from 5.0 percent to 9.7 percent. In 2003, there was a 42 percent increase in the number of periodical titles over five years earlier, despite the competition from foreign split-run magazines published mostly in the United States but containing Canadian advertising, which is now deductible by the advertisers for tax purposes.

It's interesting to note that some magazines have been doing quite well in Canada. *Reader's Digest* continues to top the list with 7.08 million readers. But *Canadian Geographic, Chatelaine,* and *Canadian Living* are in second, third, and fourth place. *Canadian Geographic* has 4.4 million readers. *Maclean's* has just under 2.75 million, and *Canadian House and Home* has over 2.5 million readers.

Mind you, paid monthly circulation figures are much lower. For the period October 1, 2004, to September 30, 2006, *Reader's Digest* was at 986,000, *Chatelaine* 645,044, and *Canadian Living* 527,694. None of the top five U.S. monthly magazines in Canada had a paid circulation of more than 374,516. Nevertheless, according to PricewaterhouseCoopers, Canadian periodicals occupy only about 15 percent of magazine-rack space, with U.S. magazines taking some 85 percent. Given the popularity of Canadian magazines, this newsstand share for them is ridiculous, surely attributable to the heavy U.S. ownership of magazine distributors in this country.

While during the past 25 years the total number of magazines purchased by Canadians fell roughly 30 percent, the average circulation for American titles fell by over 50 percent. In December 2005, the Ontario government's secretariat Magazines Canada, claimed that "roughly 50 percent of all magazine sales in Canada" can now be attributed to Canadian periodicals.

An aside. James Adams pointed out in the *Globe and Mail* on July 25, 2006, that *National Geographic*'s paid circulation in Canada had fallen from over 800,000 copies to about 374,500 in 2005, and *Playboy* had plummeted from over 500,000 to just over 61,000. What does this mean? I hesitate to speculate, but at least in the case of *Playboy* it probably has to do with widely available porn on the Internet.

How does government support for public radio and television in this country compare with the support for public broadcasting in other countries? In 2004, the OECD ranked Canada a dismal 22nd in a list of 26 countries in public funding for national public broadcasting. Not surprising. All in all, in real terms, Liberal and Conservative governments in Ottawa have chopped more than $400-million from CBC funding. Friends of Canadian Broadcasting summarizes things well:

> Our radio and television are drowning in a sea of foreign . . . mostly American . . . content.
>
> The vast majority of television programming available to Canadians during prime time hours comes from the U.S.
>
> Canadian private broadcasters, like CTV and Global, air mainly American shows during peak hours – shows that reflect American culture, concerns and ideology.

Meanwhile, 87 percent of Canadians say it's becoming more important to strengthen Canada's culture and identity, 89 percent say that the CBC helps distinguish Canada from the United States, and 89 percent say that CBC funding should be maintained or increased.

Yet in per-capita comparisons, the BBC gets more than twice as much government support as the CBC. The Scandinavians, the Japanese, and the Germans also supply far more per-capita public funding to their public broadcasters. Of the G7 nations, as a percentage of GDP, Canada provides a lower level of public support than any other nation except the United States. And while the politicians allowed the CBC's public funding to decline by $400-million over the last 15 years, the same politicians somehow expected the quality of the corporation's radio and television broadcasting to be not only sustained, but improved.

In contrast to Canada's private TV broadcasters, as the CBC's Richard Stursberg has noted,

> From 7 pm to midnight, almost every night CBC's schedule consists of Canadian comedies, dramas, documentaries, news, current affairs and sports.
>
> Canada spends the least amount of money per capita on its national public broadcaster. Meanwhile, Italy, Spain, Germany and the United Kingdom, France, Finland and others are ahead of us in spending . . . and they are hardly under the cultural pressure from U.S. that Canada is.[9]

Canada provides CBC/Radio-Canada with a poor $33 per capita in public funding. This compares with public-broadcasting support of $154 per capita in Switzerland, $124 in the United Kingdom, and in France, Germany, and Italy an average of $81 per capita. In a broader survey in 2004, among 18 major Western countries, Canada had the third lowest public funding for its public broadcaster. What's more, the survey found that "Canada's funding for public broadcasting was less than one-half of the average across the 18 western countries."[10]

To say that many Canadian radio broadcasters and arts and culture groups were shocked by the Canadian Radio-television and Telecommunications Commission (CRTC) 2005 ruling approving new satellite and digital radio services would be an understatement. Incredibly, the applicants would be required to begin their service with only 10 percent of their channels Canadian, despite the fact that private radio stations in Canada are required by the CRTC to devote 35 percent of their music to

Canadian content and, for French-language stations, 55 to 65 percent of vocal music must be performed in French.

Ian Morrison, of Friends of Canadian Broadcasting, put the CRTC decision in perspective:

> The consequences for those who want to view or listen to Canadian programs, and for Canadian artists and creators, are stunning.
>
> If the American deal is allowed to stand, it will only be a matter of time before conditions that have spawned tremendous Canadian talent, such as the requirement to play 35 per cent homegrown music, are a thing of the past.
>
> Funding for development of Canadian programming and talent will suffer as Canadian broadcasters seek to reduce the levels of investment that are several times greater than those required of the new American entrants . . . made up of nine foreign channels for every Canadian channel. . . .
>
> For more than 30 years this country has had a Canadian content policy regarding what music must be given airplay. That policy has paid huge dividends, resulting in a thriving music sector that employs thousands of people in high-wage, high-skill jobs and giving our artists the opportunity to succeed on the world stage.

Which, as almost every Canadian knows, they have very successfully done.

This said, looking at music sales per person in Canada in the first half of 2006, Canada is way down in 20th place in a list of the 26 countries with the largest music markets. While it's perhaps not surprising that the per-capita dollar value of music sales in Canada is well below that of countries such as the United States, Britain, and Japan, we're also well below countries such as Norway, Denmark, Belgium, France, Austria, Germany, and Switzerland.[11]

In 2001, Canadian music artists had 16 percent of the market share of sales in Canada. By 2004, this had increased to 25 percent. While

75 percent of all CDs, DVDs, and tapes sold in Canada are American, 93 percent of survey respondents say that music by Canadian artists is better than or equal to music by foreign artists. Sales of Canadian albums increased from 6.8 million in 2001 to 8.5 million in 2004, a 25 percent rise, while the sales of foreign artists fell over 15 percent during the same period.[12] However, in 2006, total sales of CDs, music DVDs, and other music items in Canada fell by a record 12 percent, and at this writing it appears that 2007 will have been an even worse year, due mostly to illegal Internet downloads.

Turning briefly to film, in 2005, Canadian films accounted for only 5.3 percent of tickets sold at Canadian movie theatres. In the words of Marke Andrews of the *Vancouver Sun*, "The figures reinforce the fact that Quebecers flock to French-language Canadian films, while English-speaking Canadians largely ignore home-grown films. The French-language market accounted for $36.4 million of the box office, while English-language films earned just $7.7 million."[13] According to the book *What Canadians Think,* by Darrell Bricker and John Wright of Ipsos Reid, 73 percent of Canadians cannot name a single Canadian film that they've seen during the past year.

Despite some bad broadcasting industry news, in the spring of 2005, Statistics Canada reported that "Canadians are increasingly choosing homegrown news and public affairs shows and other programming on Canadian television." Meanwhile, a surprise: the proportion of time allocated to watching sports fell on regular, pay, and specialty television.

In 2005, Canadian content on television was about 57 percent for conventional TV and about 44 percent for pay and specialty TV, the latter up from 40 percent in 1998. Of course, Francophones watch much more Canadian TV than do Anglophones, spending seven times as much time watching Canadian comedy and three times as much watching Canadian drama.

In 2003, the average Canadian spent 22 hours a week watching TV. In 1998, young men and women aged 18 to 24 averaged 31.9 hours in front of a TV set, but by 2003 this had dropped to an average of 26.6 hours.[14] Of course, a major reason is the Internet. Statistics Canada says that in

2005 some 16.8 million adult Canadians (68 percent) surfed the Internet for personal non-business reasons. About three of every five use the Internet to read news, check sports results, or do their banking online.

In 1998, only 49 percent of Canadian households had a personal computer. By 2005, that had increased to almost 75 percent, and more and more Canadians say they spend a greater amount of time online than they do watching television. In fact, since 1999, the time that Canadians spend both listening to radio and watching TV has dropped. Despite this, 419 FM radio stations pulled in almost $1.1-billion in revenue in 2006, up from $806-million in 2002. Meanwhile, the number of AM stations had fallen in 2006 to only 178, down 31 since 2002.

A word from *Globe* television columnist John Doyle about the CRTC. Private broadcasters, as always, want to cut back on their required support for Canadian programming. For Doyle,

> The vastly profitable commercial broadcasting racket in Canada doesn't need a break. It needs regulation and a sharp reminder about cultural responsibility. Later, they can cry all the way to the banks. These days, the CRTC's policy is to pamper the pampered and let everyone thrive except the creative community in Canadian TV. And yes, it's a disgrace.
>
> What's needed is more regulation, not less. What's needed is more firmness with broadcasters, not the slippery business of allowing empty promises to be made in return for permitting more overlap and concentration.[15]

Having strongly lobbied Ottawa to provide major government subsidies for the Broadcast Program Development Fund, Canada's cable television broadcasters, fat with burgeoning profits, now say that they want to keep the hundreds of millions of government dollars that they receive without turning it over to the program fund. Talk about chutzpah!

In the fall of 2005, Canada and many other countries battled the United States in a combined attempt to establish a new international convention that would allow countries the right to treat their cultural industries differently from other industries in international trade agree-ments. Canada has been a strong proponent of an effective international agreement since the late 1990s, when Sheila Copps led the charge, but it is something that the United States and its powerful culture-exporting conglomerates have fought against for decades in their desire to elimi-nate any and all barriers to U.S. cultural industry dominance.

In October 2006, 151 countries voted in support of the new agreement, while only two (the United States and Israel) voted against. The then *Toronto Star* columnist Graham Fraser wrote of Canada's efforts: "The campaign led to a remarkable coalition, not only between English and French speakers, but also between the federal and Quebec govern-ments."[16] Bravo! It was a very important victory, of which few Canadians are even aware, but it will make a huge difference in our ability to have a vibrant group of cultural industries thrive in our own country.

The Harper government, in its 2007 budget, said, "As Canadians we are proud of our history and culture and the things that make us unique." But with worrisome new developments – such as the unfortunate take-over of Alliance Atlantis Motion Picture Distribution and their massive library of Canadian films and television shows[17] (many financed by gov-ernment tax credits and grants), the remarkable decision by the CRTC to stop regulation of network advertising limits, the suggestion that Canadian television content regulations be weakened, and the growing lobby in Ottawa to remove restrictions on the foreign ownership of book publishers and bookstores – one has to question just how proud of our culture the Harper government really is.

To the dismay and astonishment of many, in the fall of 2007 the CBC secretly and incomprehensibly sold off its foreign sales division to a non-Canadian firm, without giving Canadian firms a chance to bid on the sale. So, in the future, 135 government subsidized made-in-Canada shows will be internationally conrolled by non-Canadians.

Although the Harper government certainly didn't intend that people who read their budget should consider this, clearly one of the most important things that does make us unique is just how weak our government's support for culture really is, and just how much foreign ownership and control of our country Ottawa is prepared to allow and even encourage.

The Canadian Council for the Arts has an excellent paper comparing government cultural funding in which it states, "Canada provides relatively low levels of funding for the arts, or at least stands just below the middle in terms of ranking."[18] The document shows that government arts spending in Canada, at 0.21 percent of GDP, is less than half Finland's rate of 0.47 percent. Public arts spending as a percentage of total public spending, at 0.93 percent, is far below the levels for Finland (2.10 percent) and Germany (1.79 percent).

In per-capita terms, Canadian support for the Canada Council for the Arts is only about two-thirds of the level for similar organizations in Australia and New Zealand, but is far, far below levels in the United Kingdom. Total arts grants there in 2003/2004 came to just under $23 per-capita, compared to $4.15 in Canada and only 44 cents in the United States.

Let me end this chapter on culture in Canada with four brief quotes. First, from filmmaker Atom Egoyan, regarding the takeover of Alliance Atlantis: "Culturally speaking, we'll become another [U.S.] state, because there is no incentive to continue to develop a domestic industry or a distinct alternative to the American system."[19]

Next, from Jeffrey Simpson in the *Globe,* on the state of book publishing and the heavy dominance of American publishers in the Canadian market:

> This morning's Globe and Mail bestseller non-fiction booklist shows two of 10 entries by Canadian authors.
> English-speaking Canada is the only place of any size in the world where only two of 10 bestselling books would be by writers from that country.[20]

And next, from Ian Morrison regarding the controversial sale of Alliance Atlantis: "This is a lawyered deal representing creeping American control, foreign ownership by the back door."

Lastly, from TV critic John Doyle, commenting on the new CRTC chairman's suggestion that we need "a lighter approach" to TV regulation: "Less regulation of broadcasting in Canada is an eccentric proposition. Bluntly put, there are some people who can make the reasonable point that our airwaves have been seriously polluted. We are awash in American network drama in prime time and oodles of cheesy celebrity news."[21]

Well, yes, awash we indeed are, not quite yet slipping below the surface and drowning, but there are people in Ottawa who are clearly anxious to pilfer our remaining life preservers.

THE MEDIA IN CANADA

"THE PRESS VERSUS THE PEOPLE"

"Media in Canada are the most concentrated in the world."

Lawrence Martin of the *Globe and Mail* wrote a fine column in 2005:

The media, to the tune of about 90 percent, ripped the Martin government to pieces over its decision to reject Washington's missile-defence plan. The people went the other way; they favoured the decision in polls by a 20 percent margin, which, in political terms, is a landslide.

Today's press, most strikingly on the question of U.S. relations (missile defence, Iraq, defence spending, taxation, etc.), has become concertedly conservative, moving to the right of the people.

The conservative media tend to favour a clear embrace of the United States and its values. Canadians themselves show little inclination to go that route. It is a storyline – the press versus the people – that runs right to the heart of the debate over the future of our country and to the heart of politics.

The end result is two large newspaper chains on the right, none on the left.

Meanwhile, at Maclean's, a former editor of The National Post is in charge. At Policy Options, formerly a very liberal magazine, two former employees of Mr. Mulroney run the show.

In traditionally liberal Ottawa, policy-makers wake up to four newspaper choices that all tack conservative. The largest segment of the population are centre-left Canadians.

On missile defence, the media tone was remarkably hostile. The issue was examined not so much on the basis of what Canadians think, but on what the Bush administration would think. It was as if – after 138 years of existence – we were still strapped down to a client-state mentality wherein the driving imperative was approval from a higher authority.[1]

Well-known journalist and author Geoffrey Stevens puts it this way:

Media ownership is more concentrated in Canada than in any other western country – and our laws to protect the public interest from excessive media power are the weakest anywhere.

The media have no interest in enlightening the public about the perils of entrusting too much power to too few media owners.[2]

To those who say that there is no problem, given the proliferation of radio and TV stations, Stevens says,

What use are 100 voices if they are all saying the same thing, promoting the same values, advocating the same policies?

Three times in the past 37 years, Ottawa has conducted inquiries into ownership concentration and cross-ownership (print owning broadcast and vice versa). But nothing has been done.

Stevens points out that the Southam chain of newspapers used to be known for quality coverage, especially in international news.

Then they fell into the hands of Conrad Black. After Black had done all the damage he could, he enriched himself by selling the papers to the Asper family, of Global TV. The Southam papers had 11 foreign bureaus when the Aspers acquired them. Only two remain today.[3]

What's worse, the Aspers have served notice that they intend to pull CanWest Global, their print and broadcast behemoth, out of Canadian Press, the 89-year-old national newsgathering collective. The departure may cripple CP and it will leave CanWest readers and viewers with even less news of Canada and the world.

As for CTVglobemedia Inc., owner of CTV and the *Globe and Mail*, and their takeover of CHUM Ltd., with its 33 radio stations, TV stations, and 21 specialty channels, "The takeover would never be allowed in the United States where laws against excessive concentration and cross-ownership are enforced. But not in Canada."[4]

I talked about all this to a top Ottawa media expert who asked to speak "off the record." He said only Italy has such an appalling level of media concentration. I asked how could this be in a democracy like Canada. Was it the political donations? No, though they had been a factor in the past. More to the point was the media's ability to influence votes. The right-wing publisher appoints a right-wing editor, the editor hires the journalists, and the journalists avoid offending the boss. There is an "unspoken mutuality of perspective." There develops "a serious fear" among politicians of no news coverage, or of stridently negative coverage. And, of course, there is an even greater fear among journalists with families to support, and mortgage payments to make, and kids to send to university. Don't cross Conrad Black. Don't offend the Asper brothers. "The prevailing government ideology, despite all the enquiries and reports, is 'hands off the media.'" And the media raises all the red herrings about a 500 channel universe and the thousands of blogs. So what if the news desks have been decimated? So what if so many channels feature American junk, cooking shows, sports, reality TV?

Kim Keirans, director of the School of Journalism at the University of King's College, Halifax, writes:

Canada is said to have a free press. But the three "C"s – concentration, convergence and cross-ownership are eating away its foundation.

Media in Canada are among the most concentrated in the world. In 2004, three companies controlled 63.3% of all daily newspapers in Canada.

Canada has 102 English and French daily newspapers. Only six of those newspapers remain truly independent.

Bear in mind that this is many years after the Davey Report, the excellent Kent Commission, and other government studies on media concentration, including, more recently, a 2006 report from the Senate. Almost every prediction and warning about growing media concentration in these reports has come true. As Kierans writes:

> Residents of cities such as Vancouver and provinces such as Saskatchewan, New Brunswick and Newfoundland live with "monopoly" and "multiple media" ownership. For example, in Vancouver the media company CanWest owns both daily newspapers and has a 70.6% broadcast share with its television stations. This domestic monopoly situation puts into question the role of media as an agency of democracy in the lives of Canadians; a point repeatedly made in various studies.

As others have observed, powerful corporate control of the media sharply narrows their role as critics. American media critic Ben Bagdikian points out it also often allows advertising values to dominate the news process where, as Kierans writes, "the basic business system" isn't criticized.[5]

The 2006 Senate report said, "No real democracy can function without a healthy, diverse and independent news media to inform people about the way their society works. The argument is that in a democracy, government should foster healthy and independent news media." For

Kierans, "What we see is a move from public interest to market interest. Other countries such as France, the United Kingdom, Germany and the United States, take broadcasting regulation and an independent media more seriously." In these countries, there are stringent controls limiting concentration of media control and restricting cross-ownership in broadcasting and newspapers.

For quite some time, there has been concerted pressure to allow increased foreign ownership of the media in Canada, including the telecommunications carriers and broadcasters. Whatever you may be told to the contrary, inevitably editorial and news decisions would be controlled outside of Canada.

The final report by the Senate in 2006 once again criticized the high level of concentrated ownership, convergence, and cross-ownership of the media and suggested new rules to curtail it in the future. Since the Senate report, CTVglobemedia has taken over CHUM, Rogers has purchased five CITY-TV stations, and Astral Media, which now has 81 radio stations, has taken over Standard Radio.

Peter Desbarats, of the University of Western Ontario, writes, "My own experience at competitive newspapers in Montreal, Winnipeg, and Toronto from the 1950s through the 1970s, and that of the majority of my colleagues, convinced us that competition was the *sine qua non* of a responsive and responsible press. And as competition lessened, more and more journalists found themselves muzzled." Meanwhile, "the Harper government muzzles the Press Gallery in Ottawa by cutting off information at the source and, after a few squawks last spring, the media accepts this!"[6]

For Kim Kierans, what has been happening to the media in Canada means that

> public dialogue is taking a back seat to a profit-driven business model.
>
> The Senate report bluntly blames the CRTC and the Competition Bureau for not using "the process available to them to limit concentration."

True to form, the Senate final report generated little media coverage (of course!) and hence little public discussion.

A few final words from Kim Kierans:

> If you want to find out about Canadian media, go to the business pages of your national newspapers. That's where you'll read about media mergers, stock prices and industry changes. In the past 40 years, independent newspapers, television and radio stations have been gobbled up and are part of converged, concentrated and cross-owned media conglomerates. Just consider Bell Globemedia, CanWest and Quebecor. These conglomerates own newspapers and magazines, television and radio networks, production houses, cable, satellite and Web portals. Canadian media are now big business driven not by public interest but by financial interests. Their main clients are shareholders, not viewers, readers or listeners. The results are fewer diverse sources of local information and less public dialogue which undermine the health of our democracy. A handful of locally-owned and independent media remain. They are an endangered species.
>
> The successful lobbying of private media has been at the expense of the public broadcaster, the Canadian Broadcasting Corporation. . . . The CBC suffers from unstable and declining federal funding. The [Senate's] 2006 *Final Report* recognizes "in a world of media concentration and cross-media ownership, the importance of the CBC as an alternate source of news and information programming is greater than ever." . . .
>
> Concerns about media concentration, consolidation and cross-ownership appear to be confined to the halls of academe and the Senate. The issue is not on the agenda of the public or public policy makers. So Canadians are in for bigger media and can look forward to diminished public discourse as the public agenda is issued from corporate boardrooms.

Meanwhile, contrary to many reports and to the perspective they so often present to politicians and regulatory authorities in Ottawa, Canadian newspaper owners are still doing well. In 2003, they had an operating profit margin of 15.1 percent, in 2004 it was 14.2 percent, and in 2005 13.3 percent. Operating profits in 2005 were $696-million, and advertising revenue increased 2.2 percent to almost $3.9-billion, while circulation revenues rose 5.2 percent to $871-million, despite declining circulation numbers.[7] While an increasing number of Canadians are reading their newspapers online, some 47 percent say that they read a paper every day. And, while circulation in the United States has been dropping, in Canada the overall picture is better. I can think of dozens of industries that would love to have operating profit margins similar to those in the Canadian newspaper business.

Further to Lawrence Martin's comments about the media and the American missile-defence plans, while most Canadians opposed Canada's participation in the invasion of Iraq, most Canadian newspapers supported it.

It's notable that while Canadian newspapers invariably refer to the Canadian Centre for Policy Alternatives as "left-wing" or "left-leaning," they never call the C.D. Howe Institute "right-wing," and rarely describe the far-right Fraser Institute as what it is. Meanwhile, the right-wing Institute for Research on Public Policy is supposedly a "non-partisan think tank" and, incredibly, the National Citizen's Coalition is also a "non-partisan organization."[8]

The pattern affects how news events are covered. When the important, years-in-the-making World Peace Forum was held in Vancouver in 2006, with some 5,000 people from 78 different countries in attendance, the press coverage ranged from appallingly bad to totally non-existent. The focus of the *Vancouver Sun* was a negative, ignorant attack on the forum, and the coverage by CTV and Global Television was minimal, with little or no positive comment. Highly regarded international experts from many countries who attended and participated in the forum were not interviewed and for the most part were completely ignored by the media. The range of important topics – such as the growing proliferation of nuclear weapons, the weaponization of space, the increasing dangers of terrorists acquiring nuclear weapons, and many other vitally important topics – was virtually ignored by the Canadian media.

In the preface to this book I briefly mentioned the mostly unreported secret meetings to discuss plans to further integrate Canada into the United States, a story largely ignored by the media. This being the case on such an important issue, how can we possibly respect the judgement and motives of our media owners and their editors?

Aaron Paton, the young Canmore, Alberta, journalist who helped break the story, won a prize for best story at the Canadian Newspaper Awards, while almost all of Canada's major newspapers, our two national papers, and our three television networks distinguished themselves by largely ignoring this huge story.

Five months after the top-secret Banff meetings, documents released through the United States' Freedom of Information Act included the official minutes of the meetings and plans for increased integration of Canada into the United States. In my interview with Aaron Paton I said

> If I was concerned before when I read the initial documents about those who were planning to attend, I am definitely much more concerned now. What we are looking at is an elite that is getting together to try and set an agenda for the political economy for the three North American countries. They refer to governments as "weak," and they are determined to dramatically alter the direction of the three countries, putting into place a series of policies that will very much be of benefit to big business.
>
> It's a very scary scenario and they are obviously well-funded. Here's a high-powered group of people getting together in secret and they're not interested in letting the public know what they're doing, even though it's of such enormous importance.

Is this interesting? Apparently not. Almost all of the media in Canada ignored the news story once again. But then again, consider this. If our media had reported the secret Banff meetings they would likely have had less room to keep us so very well informed about the activities of Lindsay Lohan, Britney Spears, Paris Hilton, Nicole Richie, Mel Gibson, and Anna Nicole Smith.

This is not the place for a long essay on the media in the United States, but it's worth noting that most Americans still believe Iraq had weapons of mass destruction before the United States invaded and that Iraq was directly involved with al-Qaeda in the World Trade Center attack. Moreover, a steady stream of George W. Bush White House lies created a pro-war climate that the media either contributed to or was painfully slow to counteract. As Amy and David Goodman wrote in the *Seattle Times,* "Media monopoly and militarism go hand in hand." A September 2007 poll showed that when asked whether "falsifying stories is a big problem in the U.S. news media," 62 percent of Americans agreed, while 34 percent disagreed.[9]

And so what if CBS was owned by Westinghouse and NBC by General Electric, two major weapons manufacturers producing materials for the war, a fact that seemed to be reflected in their TV support for the Iraq invasion? In the words of former TV host Phil Donahue, "There really isn't diversity in the media any more. Dissent? Forget about it."

Given what we all know now, and what many of us suspected at the time – that day after day, week after week, month after month, the Bush administration was lying to its own people, and to the world – how can you account for the fact that most major U.S. newspapers published the Bush claims on their front pages virtually unquestioned? Even Bob Woodward wrote a book that seemed to accept virtually all that his White House sources gave him. The *New York Times* asked, "How could all this have happened? How could some of the best, most fact-checked, most reputable news organizations in the English-speaking world have been so gullible? How can one explain the temporary paralysis of skepticism?"[10]

In 2002, Reporters Without Borders ranked the United States 17th of 167 countries in its press freedom index. In 2004, the United States fell to 22nd place. In 2005, it was all the way down at 44th.

For a final word on the media in the United States, let's turn to former CBS news anchor Walter Cronkite. In a keynote address at Columbia University he said that

> no longer could journalists count on their employers to pro-vide the necessary resources to expose truths that powerful politicians and special interests often did not want exposed. Instead they face rounds and rounds of job cuts and cost cuts that require them to do even more with ever less.
>
> It's not just the journalist's job at risk here. It's American democracy. It is freedom.[11]

In Canada, in the case of the secret Banff Springs Hotel meetings, or the so-called Security and Prosperity Partnership of North America, about which you will read more in the conclusion of this book, it's not just a question of a biased, blinkered, overly concentrated media, it's in fact the very survival of our country that's at stake.

I can't end this chapter without praising the many first-class journal-ists that we have in our country. And aside from those that are well-known nationally, within every community across Canada there are other hard-working perceptive writers that do not get the exposure they so often deserve. But for the media owners I have little respect.

A word about the *Globe and Mail*. I more often than not find an edi-torial or a column or two in the paper that I truly dislike and with which I strongly disagree. This said, I think the *Globe* for the most part has excellent daily national and international coverage. If only its editorial slant was more balanced.

We already have the huge Quebecor media conglomerate of newspa-pers, magazines, and television, the Rogers Communications empire of television, radio stations, and magazines, the Astral Media/Standard Broadcasting group of over 80 radio stations plus television and movie networks, the CTVglobemedia/CHUM newspaper, television, and radio group, Corus and Shaw Communications with their TV and radio assets, and CanWest Global with their huge newspaper and television networks, all mostly dominated by a plutocracy on the far-right of the political spectrum.

So where does that leave us? For Lawrence Martin, "You alter the character of a nation by changing how it sees itself. You change how it sees itself by changing the media."[12]

In Lloyd Axworthy's review of Linda McQuaig's 2007 book *Holding the Bully's Coat: Canada and the U.S. Empire,* he writes how our "Yankee cheerleaders" are given by our media "disproportionate platform time – just check out the panel lineups on our nightly news shows. To quote McQuaig, 'It is the views of the elite. . . . Their views are given an extraordinary amount of media time and space, which gives them considerable influence in shaping the debate and making palatable a neo-conservative political agenda.'"[13] How ironic that they should be advancing at full throttle here in Canada when their neo-con heroes in the United States are in such decline and disarray.

A few words about newspapers in Canada and global warming. In a superb article in the *Georgia Straight,* Mitchell Anderson wrote about how some Canadian newspapers "have been supplying the public with a steady diet of misinformation and skewed science on the critical issue of global warming."[14] Like earlier articles by deniers of any danger from tobacco, climate-change deniers, claiming that science does not support the idea that global warming is caused by humans, make a long list of claims quite unsupported by scientific evidence and in direct contradiction to international scientific studies by thousands of widely respected experts including the detailed reports of the Intergovernmental Panel on Climate Change involving more than 2,000 climate researchers from 100 countries.

Anderson pointed out that one such climate change denier, "who has not published a peer-reviewed scientific publication on climatology in more than a decade, has published no less than 39 opinion pieces and 32 letters to the editor in 24 Canadian newspapers." Given that expert after expert has described the material as mischief, unbalanced, baseless, and misleading, Anderson asks, "What's going on? Do newspaper editors not possess a phone?"

Anderson refers to Michael Campbell, a *Vancouver Sun* business columnist who happens to be B.C. premier Gordon Campbell's brother, who "regularly holds forth on climate change. Last year he scolded the scientific community: 'I see little evidence that proponents of man-made global warming know how damaging the shoddy science behind some of their claims had been to their cause.'"

As indicated earlier, like the big tobacco companies, Exxon Mobil Corporation set out to plant doubt in the public's minds. The skeptics they employed, wrote Anderson, "didn't have to bother defending their position in the scientific community because the public was the target audience. They restricted their pugilism to the popular press rather than peer-reviewed scientific journals."

The American Petroleum Institute called for

a campaign to recruit a cadre of scientists who share the industry's views on climate science and to train them in public relations so they can convince journalists, politicians and the public that the risk of global warming is too uncertain to justify.

If there is a significant difference between the PR efforts of the tobacco industry and the fossil-fuel industry, it is size. The oil, gas, and coal sectors make Big Tobacco seem positively puny by comparison.

Has this campaign against climate science been successful? You bet. It may well go down as the most audacious, successful and cynical campaign in public-relations history.

The result? The public is being misinformed on climate science by poor journalists that continue to tell both sides of the story even when there is no other side. The resultant political inaction might kill the planet.

Consider these numbers. A McAllister Opinion Research

poll from the fall of last year showed that fully 50 percent of Canadians still believed that "most scientists disagreed with each other about whether global warming was happening." In the U.S., the numbers are even worse. An ABC News poll last year showed that 64 percent of Americans believed the majority of scientists are still arguing about whether or not global warming was even happening.

To be fair, a great many Canadian journalists have been doing an excellent job of warning of the dangers of greenhouse gas emissions and climate change. That some Canadian newspaper editors choose not to pick up the phone and check with the scores of respected Canadian experts before they publish the opinions of very questionable sources is regrettable.

At this writing, there are some suggestions that Ottawa is considering new restrictions on media concentration, mergers, and cross-media ownership. But as is to be expected, the media owners are howling that any such actions would be "absolutely irresponsible."[15]

Professor Marc Raboy specializes in media policy at McGill University. In his opinion, "The horse left the barn a long time ago but we keep seeing more extreme cases of media consolidation."[16]

It's never too late, but what needs to be done quickly will require strong political leadership that seems to be non-existent in our country today. Meanwhile, the Stephen Harper government, in unprecedented ways, has to the best of its ability cut off the flow of information to the media and hence to the public. Most ministers are not allowed to talk freely to the media, so there has been less public comment from the Cabinet than at any time in the history of our country. The public flow of information and requests under the Access to Information Act have been slowed in a manner never seen before. And worse, much of the news media seems increasingly intimidated by the all-powerful Prime Minister's Office and are prepared to accept that they must be on a pre-approved list before they are even allowed to ask questions.

PART SIX

FOREIGN INVESTMENT, FOREIGN OWNERSHIP, FOREIGN CONTROL

"Canadians are behaving like naive Boy Scouts by failing to emulate other countries that unabashedly protect their vital economic sectors from foreign ownership."

"This is economic suicide."

In the late 1960s and throughout the first half of the 1970s, Canadians became increasingly concerned about the already high and rapidly increasing level of foreign ownership in Canada, which had reached over one-third of all non-financial industry corporate assets and over 37.4 percent of all revenues.

This concern led to the formation of the Committee for an Independent Canada (CIC), the Watkins Report and the Gray Report, and a steady stream of public opinion polls which reflected growing Canadian unease regarding the issue. After being presented with a 176,000-name petition by the CIC, Pierre Trudeau and his government brought in the Foreign Investment Review Act in 1975. In a decade, foreign control dropped all the way down to 21.4 percent. This was still very high compared to other industrialized countries, but at least it was decreasing instead of continuing to increase at an alarming rate.

After Brian Mulroney abolished the Foreign Investment Review Agency in 1986 and replaced it with the rubber-stamp Investment Canada, foreign control and the foreign share of assets and revenues began to increase once again. By 2000, the foreign control of non-financial industries was back at about the same level as it had been in the mid-1960s. The latest official Statistics Canada figures show that in 2005 the foreign share of corporation revenues was back up to 30 percent, the highest level in 30 years, and the foreign share of profits was up to 30.5

percent. Foreign companies were taking well over half of all manufacturing revenues and oil and gas revenues.

Now, in 2008, we are on our way to passing the levels that caused such great concern in the 1970s, and are rapidly proceeding to new record levels, although official Statistics Canada figures won't be available until 2009 or 2010.

In this respect, it's interesting to note that in compiling its figures, Statistics Canada does not consider companies such as Air Canada, the Canadian National Railway Co. (which held its AGM in the United States in 2006 for the first time in history), Petro-Canada, or Canada's largest oil and gas producer, EnCana, in its foreign ownership calculations, even though all four (and dozens of other important "Canadian" corporations) are already majority foreign-owned, mostly by Americans. Many other countries consider that as little as 10 percent foreign ownership can, and often will, represent effective foreign control.

For those right-wing continentalists and their comprador colleagues who make their perpetual pleas for more foreign direct investment (the Canadian Chamber of Commerce, the Conference Board of Canada, the Canadian Council of Chief Executives, the C.D. Howe and Fraser Institutes, not to mention many of our leading newspapers), we should have nothing but contempt. As we shall shortly see, what they are asking for is plain and simple: more foreign ownership and thus more foreign control of our resources, our industry, our high-tech companies, and many other businesses. How so?

Let's look at the startling figures for foreign and direct investment since Brian Mulroney declared Canada "open for business" and dumped the Foreign Investment Review Agency. From June 30, 1985, to the end of 2007, 10,807 Canadian companies were taken over by non-resident-controlled corporations. The total dollar amount of all foreign direct investment monitored by Investment Canada was an enormous $834.86-billion. Of this amount, an astonishing 97.7 percent was for takeovers. Only a truly pathetic 2.3 percent was for the hoped-for new business investment.[1]

Note these numbers well and then consider these words from a *Globe and Mail* editorial (June 21, 2007) endorsing even more foreign direct investment: "Foreign direct investment is typically welcomed when it involves new capital investment, such as creating a manufacturing plant from scratch in Canada bringing industry and jobs to the country."

So, then, three big cheers for that tiny 2.3 percent of all foreign direct investment in Canada that was for new business investment during the last 22 years.

Since Investment Canada began keeping track in 1985, some 62 percent of these foreign direct investments has been attributed to American firms. Far behind in second place is the United Kingdom at 9 percent. So when we talk about foreign ownership and control in Canada, it's predominantly American. And contrary to all the nonsense in our newspapers about Canadian direct investment in the United States exceeding U.S. direct investment in Canada, at the end of 2006 American direct investment here was $50.1-billion higher, and, more importantly, it represented a very much greater percentage of total assets and GDP.

As might be expected, a large percentage of Canadian direct investment abroad was by our good old patriotic Canadian banks, 42 percent to be exact. You know, the very guys who fund the likes of the continentalist C.D. Howe and Fraser Institutes, the Conference Board of Canada, and the Canadian Council of Chief Executives.

For some very good reasons, most Americans think they have the right to buy up as much ownership and control of Canada as they wish. For some truly bizarre reasons, many of our leading politicians, business leaders, and journalists see no problem with that. In fact, many of our political leaders and a large number of our most prominent editorial writers and columnists encourage more U.S. direct investment every time the topic comes up, seemingly ignorant of the fact that what they are asking for is even more foreign ownership and foreign control of our country.

What is remarkable is that the corporate con artists from the Business Council on National Issues who claimed that the FTA and NAFTA would encourage foreign companies to invest in Canada, because, with our supposed guaranteed access to the United States, Canada could provide a launching platform to the U.S. market, are essentially the same crew who, as the Canadian Council of Chief Executives (CCCE), are now constantly whining that Canada's share of foreign direct investment has been falling.

Is Canada running short of foreign direct investment? Hardly. In October 2006, an analysis by BMO Capital Markets said, "Inward investment is currently running well above long-run trends as a share of GDP. Indeed, strong foreign investment flows have played a role in driving the Canadian dollar above its fair value."

Not enough foreign investment? What utter nonsense. In 2006, foreign direct investment in Canada amounted to a huge $78.3-billion, the second highest amount in our history. In the first six months of 2007, foreign direct investment in Canada was the second highest amount ever for the first six months of a year.[2]

In comparison to the Canadian figure of $78.3-billion for 2006, foreign direct investment in all of the huge country of China was only about $83-billion, and in the giant U.S. economy it amounted to only $109-billion. Adjusted for the relative size of these economies, the 2006 takeover figure for Canada is six times greater than in the United States. In 2006, Canada attracted $2.12-billion per million population, France $1.32-billion, the United States only $0.59-billion.[3] And new OECD figures show that in 2007 Canada had the fifth-largest foreign direct investment inflow of any country, regardless of size.

Contrast this with the silly claims of the Conference Board of Canada and their head, Anne Golden, who told CBC Radio, "We're losing out on our share of foreign direct investment,"[4] and the *Financial Post*, "On the ability to attract foreign investment, Canada gets a D grade. While emerging economies such as China and India are becoming increasingly attractive foreign investment destinations, Canada is losing ground."[5]

But wait a minute. In 2006, foreign direct investment in India was only $8.8-billion, in all of Russia it was only $31-billion, in Mexico and Brazil less than $21-billion. Again, compare that with Canada's $78.3-billion.

What will be the impact on Canadian long-term productivity from increasing foreign takeovers? According to BMO, "One area that often suffers is research, technology, design and development."

Hilarious yet again is big business in Canada yelping almost daily about what a poor place Canada has been to invest in. Corporate taxes will have to be slashed even further. Our regulations and standards must now be harmonized with American regulations and standards, etc., etc., or else we'll be in the poorhouse.

Funny, in March 2005, the Economist-Intelligence Unit, in their *Assessing the General Business Environment,* said that Canada ranks second in the world in forecasts from 2005 to 2009 as a place to invest. Perhaps they meant the second-easiest place in the world to take over good companies.

Meanwhile, the boys at the CCCE are shoving their piles of money out of the country just as enthusiastically as they shout "We must cut taxes!"

Incredible as it may seem, given the already very high levels of foreign ownership and control in Canada, a "secret" February 2006 Industry Canada document proposed that Ottawa should encourage even more takeovers of Canadian businesses. The document called current levels of foreign ownership and control "reasonable" and suggested reducing or removing restrictions on foreign takeovers in telecommunications, broadcasting, publishing, and banking. The public servants who authored this absurd document should have been fired long ago. However, it should be noted that many key pages of the documents prepared for the Conservative Industry minister at the time, Maxime Bernier, were heavily censored. Now why do you think that would be? Could it be because the censored material was information about why more foreign ownership would be unwise, information that Bernier did not want Canadians to see?

Bernier is a former important member of the right-wing Montreal Economic Institute think tank and seems to top even the enthusiasm of Thomas d'Aquino, head the Canadian Council of Chief Executives, for selling off even more of our country. Mind you, that's difficult. In January, 2008, Mr. d'Aquino told Canadians not to fear all the foreign takeovers. He advised us that government should not impose any new restrictions; on the contrary, most restrictions on selling off the country should be removed.

Meanwhile, true to form, despite the recent huge amounts of new foreign investment and takeovers, the C.D. Howe Institute claims that "there are tax-related obstacles when it comes to foreign investment in Canada's private equity sector."[6] Apparently, a great many very enthusiastic foreign investors are not familiar with these "tax-related obstacles."

And despite the already high levels of foreign ownership and control in Canada, and the enormous new foreign direct investment in 2006, Neil Reynolds of the *Globe and Mail* tells his readers that "Canada does not have a conducive environment for foreign investment," while Ian Russell, the head of the Investment Industry Association of Canada, somehow ignoring Canada's consecutive years of record-breaking corporate profits, tells his Toronto luncheon audience, "There is too little incentive for Canadians to invest and keep investing."[7] Good grief!

Never underestimate the enthusiasm of the Harper government for selling off the part of Canada that isn't already foreign-owned. At this writing, Trade Minister David Emerson and Finance Minister Jim Flaherty have been in Beijing trying to convince the Chinese to come in and buy up whatever the Americans and Europeans don't already own. When Emerson was Industry minister in the Paul Martin government, he replied to a citizen concerned about the growing level of foreign ownership and foreign control in Canada with this profound observation: "I strongly believe that the well-being of Canada's petroleum and manufacturing industries is very much in the national interest." Then, apparently not knowing that both these industries were already majority foreign-owned and foreign-controlled, he continued, "A key element to supporting these industries is to allow and indeed promote foreign investments. . . . I see no reason at this point to introduce specific foreign ownership limits." (In 2005, before the huge takeovers of 2006 and 2007, foreign corporations already took in 56.5 percent of all manufacturing profits and 55.2 percent of all oil and gas revenue.) And then, the confidence-destroying clanger from Emerson: "Please rest assured that the investment review process is rigorous."

According to Stephen Harper, speaking in the House of Commons, foreign takeovers will only be approved if they produce a "net benefit" after being reviewed by Investment Canada and if the big foreign companies "pay their fair share of tax in this country."[8] What a joke!

Eric Reguly, writing in the *Globe and Mail* (August 11, 2006), ridiculed Investment Canada, calling it "a total pushover," which raises the question: "What is the point of Investment Canada? If it's a perpetual rubber-stamp machine that is useless, why not save the taxpayer a few bucks and sink it in the Ottawa River?"

Any suggestion that takeovers are not allowed by Investment Canada unless they provide a "net benefit" to Canada is absurd. A *Toronto Star* editorial on July 29, 2007, had it right: "Whether a takeover confers a net benefit is beside the point if that benefit is not greater than the one Canada would realize were the firm to continue operating under Canadian control . . . the net benefit test is the weakest one conceivable."

Whereas other countries around the world frequently reject takeovers if they are not clearly beneficial, Canada has few barriers. Investment Canada, since its inception in 1985, has not turned down one single takeover. *Not one, in more than 10,500!* Regulations that other countries routinely employ to retain control of their corporations are mostly absent in Canada. Yet Maxime Bernier's director of parliamentary affairs, Darren Cunningham, claimed that Investment Canada protects Canada's best interests and "bad deals will be caught."[9] (From June 30, 1985, to June 30, 2007, only 12.5 percent of takeovers in Canada were even reviewed by Investment Canada.)

Economist Mel Watkins says Investment Canada has given a whole new meaning to "Buy Canadian."

As to Harper's promise that big foreign companies will pay their fair share of taxes in this country, Leonard Farber, of Ogilvy Renault, writes in the *Globe and Mail*:

A former Finance official says "foreign takeovers will result in less tax at both the federal and provincial levels."

First, he said, there'd be no more tax revenue from Canadian investors. "[Also], private equity will load up debt in Canada . . . rendering the operation basically non-taxable in Canada, and interest crossing the border will be free of withholding tax."[10]

By the end of 2006, the following industries in Canada were majority or heavily foreign-owned: manufacturing, the petroleum industry, chemicals and chemical products, mineral fuels, non-metallic mineral products, food processing and packaging, electric products, tobacco products, machinery, transportation equipment, computers, major advertising firms, meat packing, brewing, aircraft, etc., etc.

Altogether, some 36 different sectors of the Canadian economy are now heavily or majority foreign-owned and/or controlled. And now the Harper government is under increasing pressure to allow the foreign takeovers of Canadian utilities, airlines, bookstores and book publishers, telecommunication companies, and other industries.

In comparison, in the United States there's *not one single industry* that is majority foreign-owned or controlled. Not one! And only two have foreign ownership of assets in the 30 percent range.

Another way of comparing foreign direct investment in Canada and the United States is as a percentage of GDP. In Canada in the 1990s, it averaged 22 percent. In the United States, it was only 8 percent. At the end of 2005, only 4.7 percent of private industry in the United States was controlled by foreign firms. As already indicated, over half of all manufacturing in Canada is now foreign-owned. In comparison, among the other 29 OECD countries, all of the following are far below the level in Canada: Japan, Germany, the United States, Poland, Norway, Italy, the Netherlands, Finland, the United Kingdom, France, Sweden, and the Czech Republic. No other major industrialized country has a level of foreign ownership of its manufacturing anywhere near as high as Canada's.

In the decade of the 1990s, foreign direct investment in Canada amounted to $126-billion. By 2000, it was over two and a half times as much as it was in 1990. By the end of 2007, Investment Canada had recorded new foreign direct investment in takeovers alone of $603.88-billion.

It's always interesting to ask our gung-ho sellouts just how much of the country they're prepared to sell off. None of them will ever give you an answer. I asked Anne Golden of the Conference Board of Canada this question, and all she could say was, "That's an interesting question." Try writing Stephen Harper a letter asking him this question and see what reply you get.

If you have a strong stomach, go to the Investment Canada website (http://strategis.ic.gc.ca/epic/site/ica-lic.nsf/en/h_lk00014e.html) and have a look at any one month of takeovers of businesses in Canada. Month after month, year after year, in every region of the country, the long list of takeovers is appalling: petroleum and mining companies, forestry and energy distribution companies, clothing and design companies, computer and software companies, wholesale and retail operations, hotels and resorts, a multitude of important service industry companies, our largest steel producers, insurance and finance firms, real estate and construction companies, home heating and power companies, asset management firms, restaurants, breweries, bakeries, research firms – the list, month after month, goes on and on and on.

It's remarkable but too sad to be laughable to hear the constant whining about poor productivity, lack of Canadian patents and innovation, poor levels of high-tech exports, and so on, when almost every week another Canadian high-tech company is taken over, by a foreign corporation. As others have pointed out, 125 such companies in the Ottawa area alone were taken over in the decade ending in 2003. And now you can also wave goodbye to the Canadarm and Radarsat-2, along with all the other high-tech companies recently taken over by Microsoft and IBM.

In the late 1990s, there were over 40 large Canadian petroleum companies. Since then, foreign companies have purchased most of them. Now there are only six left.

One really must wonder what the level of foreign ownership has to be before Mr. Harper and his corporate colleagues are satisfied that enough is enough. Sixty percent? Seventy? Eighty? Or should it be 100 percent foreign ownership and control? Too bad that some Member of Parliament or the Press Gallery hasn't long ago asked such a question of our political leaders.

Today, the goal of Investment Canada is to *facilitate and solicit even more foreign direct investment,* not to limit or control it. This was the goal of the Mulroney government when it abolished the Foreign Investment Review Agency, and both the Chrétien and Martin governments enthusiastically continued this policy. If anything, the Harper government, despite some rhetoric to the contrary, seems eager to open the door even wider and allow new record levels of takeovers of our businesses, resources, and land.

At the same time, it's interesting to note that the Canadian public has consistently shown that it wants otherwise. Year after year, in poll after poll, Canadians say we already have too much foreign ownership and control and we don't need more. In September 2007, a Canadian Press/Decima poll said that 72 percent of Canadians reported that they wanted the federal government to do more to limit foreign takeovers, including 66 percent of Conservatives polled.

In sharp contrast, but certainly true to form, by a margin of eight to one, Canadian CEOs said they were opposed to any controls,[11] while the likes of Roger Martin of the Rotman School of Management and the Royal Bank's Gordon Nixon warn against any "harsh" action or "overly aggressive negotiations by Investment Canada" to stem the tide of foreign takeovers,[12] and the Canadian Chamber of Commerce has warned the Harper government to interfere as little as possible in future foreign investment takeovers.[13] If that doesn't tell you a great deal about our Canadian business elite, I don't know what will.

The conventional wisdom among our corporate elite, much of our media, and our federal and provincial politicians would have us believe that the development of Canada and our standard of living has been largely due to the influx of foreign capital. Not so. In fact, most of the massive takeover of corporations in Canada has been financed by our own good old reliable Canadian banks and our other financial institutions, including the caisses populaires, and our very own pension funds.

Canada's leading investment bankers and our largest pension funds should be called before a parliamentary committee to explain why they devote so much of their capital to financing the foreign acquisition of Canadian assets. Instead of financing the sell-off of our country, why aren't they financing the growth of Canadian companies that need expansion capital? Companies that are bought up don't create many new jobs, if they create any at all, and they rarely grow into globally competitive firms. *Toronto Star* economics columnist David Crane puts it well: "Our investment banks and pension funds are not aligned with the long-term interests of Canada, and that has to change," instead of making their priority their big commissions and capital gains from selling off the country.

Are we short of money in Canada? Hardly. Aside from the ever-growing assets of our big banks, our four largest public-sector retirement funds had $420-billion in assets by spring 2007. And, as *Globe and Mail* economics reporter Heather Scoffield has pointed out, "Canadian corporate liquidity – i.e., money in the bank – has risen steadily over the past 15 years, and is now so far in record territory that historical comparisons are almost meaningless."[14]

Worth noting is that Statistics Canada balance-of-payments figures do not include the massive amount of funds foreign corporations raise in Canada every year, so that the actual value of foreign ownership is substantially higher than the published balance-of-payment figures. For example, the takeover of the recreational products division of Bombardier was financed by two Canadian banks (BMO and RBC) and a Quebec caisse. The CIBC was the leading lender in the takeover of Shoppers Drug Mart. The CIBC and the Bank of Nova Scotia helped finance the Yellow Pages sale. A full list would fill many pages.

As I have pointed out many times in the past, no one in Ottawa knows just how much of the sale of our country has been financed with our own money, not the Department of Finance, not the Bank of Canada, not Statistics Canada, not the Prime Minister's Office or the Privy Council Office – no one!

Why is that? Simple. None of them are interested. They don't care. One thing is for sure: such appalling ignorance could not be found in any other developed country.

Meanwhile, incredibly, thanks mostly to our banks and other Canadian financial institutions, the outflow of foreign direct investment from Canada in the period from 1995 to 2004 was greater than the outflow from Germany, the United States, Italy, Finland, Sweden, Portugal, Belgium, Luxembourg, Norway, Austria, New Zealand, Australia, and Ireland put together.

Under the terms of both the FTA and NAFTA, Canada gave away many of the tools used by nations around the world to keep a reasonable check on excessive and/or detrimental foreign ownership and control, and even abandoned many of the options to ensure that takeovers had to clearly bring benefits to this country.

All of this raises two very interesting and important questions, one easy to answer, the other very difficult. The easy question is why do other developed countries reject such high levels of foreign ownership and foreign control?

There are many important reasons, far too many to do justice to them in this chapter, but to begin, here's just one. Foreign subsidiaries import much of their goods and services from their parent company, almost always at high, non-arms-length prices. Here are G7 figures for imports of goods and services as a percentage of GDP.

Canada	41%
Germany	28%
United Kingdom	27%
France	24%
Italy	24%
United States	13%
Japan	9%

These figures represent a huge loss of overall economic activity. They are caused by excessive imports resulting from excessive foreign ownership and control. Overall, foreign firms operating in Canada import three times as many parts and components and services as similar sized Canadian companies. In a truly remarkable comparison, an OECD study showed that the ratio of foreign parts and components in manufacturing in the United States was 13 percent, in Japan 7 percent, while in Canada it was over 50 percent.

Very high levels of imports of goods and services purchased by foreign branch plants at inflated prices from their parent companies have been very profitable for the foreign parent. The loss of profit for the branch plant is equally intentional, as it neatly reduces taxes payable in Canada. This will likely continue to be true even if tax rates are now lower in Canada, as Eric Reguly explains below.

Again, how does our level of import penetration in Canada compare with other developed countries? In 2005, the OECD 30-country average was only 20.2 percent. No wonder both the United States and Japan have long had such very low unemployment rates!

The *Globe*'s Eric Reguly adds his similar viewpoint on the impact of foreign takeovers:

> There's something wrong with this picture. Canada is a G7 country that doesn't act like one. We are told that Canada is no worse for it, that foreign companies can treat their Canadian subsidiaries with love and care, that some of them even expand in Canada and add jobs. But, how many jobs would the companies have created if they were still owned and managed locally? And ask the lawyers at Osler, Hoskin and Harcourt how they feel about Inco, one of the firm's biggest clients, going to an owner in Rio de Janeiro.
>
> The loss of Canadian head offices is an issue for all Canadians, if only because foreign owners restructure their foreign subsidiaries to minimize taxes. Watch the tax payments made by Inco and Falconbridge plummet as they become branch plants. Typically, the Canadian subsidiaries of foreign companies are loaded up with debt. Debt payments are tax deductible. Ottawa's corporate tax haul can only fall. Individual taxpayers will have to make up the difference.[15]

And, of course, provincial tax revenue will also fall.

Of the 250 largest private companies in Canada, over 54 percent are foreign-controlled. For reasons that are entirely impossible to comprehend, the Liberal government abolished the previous long-standing requirement for these corporations to publicly report such information as profits, assets, and return on capital. Bizarre, to say the least. Of the top 50 foreign-controlled companies in Canada, 44 are 100 percent foreign-owned.

Among the long list of foreign companies in Canada that are 100 percent foreign-owned are: General Motors, Chrysler, Ford, Honda, Costco, IBM, Safeway, Cargill, McDonalds, General Motors Acceptance Corp, Nissan, Mitsui, Kinder Morgan, Siemans, Mazda, Unilever, American Express, Sony, and Citigroup.

Three points. First, many of these companies are 100 percent foreign-owned because they don't want to share corporate information, especially matters relating to transfer pricing and debt charges, with nosy Canadian shareholders who would object to profits being transferred out of Canada.

Second, even if you wanted to buy shares in their "Canadian" companies, you couldn't, unless you wanted to buy shares in the foreign parent, if they are available. And, of course, you would likely buy these shares in the United States, so more Canadian investment dollars would end up south of the border.

Next, for Canadian investors the choice in Canada is narrowed as we sell off the ownership of our companies to foreigners. And, at the same time, the resulting smaller range of TSX alternatives artificially inflates the price of Canadian shares.

In 2005, the outflow of dividends, mostly to the United States, was at a record level, $15.46-billion, up from $12.14-billion in 2004. In 2006, they increased again, to $18.18-billion.

Meanwhile, the *Globe and Mail*, in a lead editorial (August 20, 2006), called foreign ownership of Canadian corporations beneficial and supported the idea that it provides a net benefit to Canada. The *Toronto Star* was far more perceptive, pointing out that crucial decisions concerning business in Canada will be made in corporate boardrooms "by men and women with no ties to this country," while the branch plants and subsidiaries cannot plan their own affairs and rarely are allowed to compete with their parent companies or determine where they purchase their parts, components, and services.

As Professor Harry Arthurs, one of Canada's leading labour law scholars, has observed about the head offices that remain in Canada, "That's what the whole thing is about. That is why it's called hollowing out. You still have something that looks like a head office but it's hollow – there's nothing inside. The mere number of head offices means nothing."

Ian Telfer, chairman of Goldcorp, says, "It's tragic that Canada has lost so many potential champions with such a devastating effect on support industries such as accountants, bankers, geologists and a knowledge base that starts to disappear."[16]

Tom Caldwell, of Caldwell Securities, in powerful full-page ads in the *Globe and Mail* and the *National Post* on July 27, 2007, under the heading "The Sellout of Corporate Canada," wrote, "The loss of head offices and industrial leadership by Canada is one of the great corporate tragedies of our time. Future generations of Canadians, wishing to climb the corporate ladder, will increasingly be compelled to go elsewhere. The current trend guarantees Canada's losing some of its best and brightest people."

For Gerald Schwartz, CEO of Onex Corp.,

> What happens when companies headquartered in Canada increasingly are being sold to foreign head office buyers?
>
> For one thing, the ambition of Canada's young business people is stopped dead in its tracks. They have to move to other countries, usually the United States, to achieve their aspirations.
>
> When a head office leaves Canada, so do all the support positions at accounting firms, law firms, recruiters and other key suppliers of head office intellectual capital. Just watch the corporate donations budgets decline when a Canadian company gets taken over by a foreigner.[17]

Suggestions by the Conference Board and others that foreign companies create more head office employment in Canada is ludicrous. A

head office is a place where vital decisions are made, not a place mis-
leadingly designated in this way because it has permission to order its
own toilet paper.

How do other countries handle this challenge? In the United States,
when China National Offshore Oil Corporation tried to take over the
American oil firm Unocal Corp., the ninth largest U.S. oil company,
Washington stepped in to thwart the Chinese bid. In May 2007, the
Australian government regulator blocked the attempted takeover of
Quantas Airways.

In Italy, "there is political resistance to foreign ownership" as wit-
nessed by failed attempts to take over a large Italian telecom company.
Selling control of telecoms in both France and Germany to outsiders is
"all but unthinkable and practically speaking impossible."[18]

In France, when Pepsi attempted a takeover of the famous Danone
food company, the French prime minister, Dominique de Villepin,
warned Pepsi not to proceed further.

In both Germany and Spain, the government has made it clear that it
prefers domestic ownership of major corporations, and in both Italy and
Poland, government is discouraging foreign attempts to take over banks
and other firms. In France, there is legislated prohibition of foreign
takeovers in 11 different major sectors of the economy. Germany, India,
Japan, Russia, and Bolivia are all in the process of toughening their regulations regarding foreign ownership.

In the European Union, new EU takeover rules now allow countries
a wide variety of options to turn back foreign takeovers. In Indonesia,
there are tough new rules to thwart foreign takeovers. And in the fall of
2007, Japan took steps to make foreign takeovers more difficult. There
are many, many other examples every year of governments stepping in to
control foreign takeovers in every part of the world.

One reason excessive foreign ownership is discouraged in other countries is that the dominance of foreign corporations in an industrial sector
inevitably brings pressure on government policy in both domestic and
foreign policy development. The job of foreign subsidiaries is to make as
much profit as possible for their foreign parent. The Canadian national

interest plays no role in such considerations. Exxon, to give one example, tells Imperial Oil what to do about maximizing or minimizing their public positions regarding petroleum reserves, and the result no doubt benefits Exxon – but it may well not be in Canada's national interest.

Other downsides include the marginalization of Canadian directors, the inability of subsidiaries to compete with their parent in export markets unless permission is granted, and, above all, the fact that key decisions on the opening and closing of plants, the level of wages and dividends, and the adoption of U.S. standards, values, and policies are made by the foreign parent.

There's no room here to do proper justice to all the other foreign ownership negatives, but much can be summed up in my favourite quote on the subject of takeovers, which comes from none other than Brian Mulroney: "I've yet to see a takeover that has created a single job, except of course for lawyers and accountants."[19]

Try looking at the number of jobs per million dollars in sales and compare Canadian firms and U.S. subsidiaries where their numbers are available. The numbers are shocking and most revealing. In 2000, foreign firms in Canada made 53 percent of all manufacturing shipments in this country, but employed under 32 percent of manufacturing workers.

The second important question referred to above is much more difficult to answer, and truly borders on the bizarre. After the takeover of the Hudson's Bay Company and Dofasco in February 2006, *The Economist* had this to say: "In many other countries, the sale of national heirlooms would spark fierce opposition. Not in Canada."

Peter C. Newman says, "In all other developed countries the economic elite defend their country's sovereignty because not only is it in their own interest to do so, but they are proud of their country and wish it to be more than a place where their children and grandchildren can best look forward to being serfs."

David Crane pinpoints one element of the problem: "The upsurge in foreign ownership and control in the Canadian economy would not be taking place if our financial markets were focused on building Canadian companies, rather than selling them."

For Thomas Caldwell, one of the key problems has been that, "increasingly, power accrued to corporate managers who saw themselves as elite, entitled not only to inflated compensation as hired hands, but also to the rewards of ownership without risk [while] consultants continually justified wildly inflated pay, bonuses and option schemes for senior management. To hear those managers described as risk-takers is a joke."

It would take a whole book to explain properly why our country is being sold off. Yes, some, or much, or even most, of it is related to greed, pure and simple, but that alone cannot explain the extraordinary and virtually unique absence of patriotism and loyalty to homeland among so many of our corporate establishment. Surely, though, the fact that so much of our media is either American or controlled by our own far-right conservative continentalists is one reason why our hollowed-out country is sleepwalking back to colonial status.

As discussed above, for years now, Canada's business community, always promptly echoed by our business press, has been wailing about our supposed low levels of foreign direct investment. As indicated earlier, a good way to measure foreign direct investment is as a percentage of GDP. By this measure, in 2005, foreign direct investment in Canada was higher than in 18 OECD countries, including France, Finland, Germany, Poland, Mexico, Austria, Norway, the United States, Turkey, Greece, Italy, Iceland, and Korea. It was 3.5 times as high as in the United States, 51 times as high as in Japan, and over a third higher than the OECD average. It was also higher than in non-OECD countries such as South Africa, Brazil, Russia, and India.[20] And this was *before* the huge takeovers in 2006 and 2007.

It's worth repeating that much of so-called inward foreign direct investment in Canada is, in fact, *not* inward at all. Rather, it is money generously supplied by Canada's banks and other financial institutions to finance the sell-off of Canadian businesses to foreign, mostly American, buyers.

In June 2006, a new Statistics Canada study was published (*The Daily*, June 2) that showed, among other things, that foreign-controlled corporations already take 30 percent of all corporate operating revenues in Canada, and their assets were already over $1.1-trillion. In the words of Statistics Canada, in 2004, "foreign-controlled profits soared to a staggering record of $68 billion." For Statistics Canada to use a word like "staggering" is highly unusual. Yet, typically, there was barely a tiny ripple in the media.

In 2005, world inflows of foreign direct investment were almost $916-billion (U.S.). China had an inflow of about $60-billion, all of Africa $30-billion, Russia about $25-billion, Mexico about $17-billion, India about $6-billion. Canada, with only 32.7 million people, had had an inflow of $41-billion, the third highest annual amount in our history.

During the 10 years from 1996 to 2005, foreign direct investment in Canada totalled $330.14-billion. In the previous 10 years, it amounted to only $77.25-billion. And, as mentioned earlier, in 2006, foreign direct investment in Canada, at $75.6-billion, was the second highest total in our history.

Here are just a few of the 26 countries that in 2006 had foreign direct investment inflows below ours: Hong Kong, Russia, Brazil, Mexico, Chile, India, South Korea, South Africa, Argentina, Taiwan, Peru, and Saudi Arabia. Of a list of 26 OECD countries, 21 had inflows of less than $10-billion (U.S.).

Of course, foreign-controlled assets are much greater than foreign direct investment. As mentioned earlier, you can easily control a widely held corporation worth $100-million with a $10-million direct investment. Another important way that the value of foreign takeovers is understated is the fact that the investment review division of Industry Canada reports acquisitions by book value, not market value. Nor does it adequately reflect two of the largest sources of the expansion of foreign assets: funds raised in Canada for acquisitions (for example, the takeovers of the Montreal Canadians, Teleglobe, Shoppers Drug Mart, and so on), and expansion by the use of retained earnings. The result is very misleading numbers that downplay the situation.

In May 2006, however, Statistics Canada released a shocking new report. I read six newspapers a day and saw not one single word about the document, titled *Market Value of Foreign Direct Investment Position.*[21]

The report showed that while the oft-reported book value of foreign direct investment in Canada at the end of 2005 was $415.56-billion, the true market value of that investment was an enormous $771.33-billion – an increase of $355.77-billion greater than any previously reported figure! Moreover, the market value of foreign direct investment in Canada had increased by a huge 47 percent between 2002 and 2005.

This unreported study came just eight days after the *Globe and Mail* said Canada should make it even easier for foreign companies to buy up ownership and control of Canadian companies. This despite the fact that Canada already had the highest degree of foreign ownership and control of any major developed country, and that foreign ownership and control has been increasing now for 18 consecutive years.

Typical of the constant corporate wail that we're not getting enough foreign investment was an August 9, 2005, *National Post* piece by Jack Mintz, then president and CEO of the far-right continentalist C.D. Howe Institute. "Canada has experienced a loss in the worldwide share of foreign direct investment. . . . [A main reason is] an onerous tax regime . . . [and] withholding taxes levied on dividends and interest payments made to non-resident owners."

More recently, Mintz said there's nothing to be concerned about, since "Canada historically has not been the place where people are particularly trying to buy assets."[22]

Interesting. If Canada has been such a poor place for foreigners to invest in, then why do they keep setting new records buying up more and more of our country, year after year, with over 12,000 companies already under their control?

The problem in Canada isn't that we need more foreign direct investment. The problem is that we already have far, far too much foreign ownership and control. The last thing we need is more of the same.

One thing I continue to find disconcerting is the persistent and widespread misinformation in the press about foreign ownership and control. Whether it's the number of Canadian companies taken over or the number of Canadian takeovers outside of Canada, or the amount of R&D foreign companies do in Canada, or the dollar value of takeovers, in far too many cases the information the public receives is simply not true and is frequently terribly misleading. The OECD noted this in June 2007, when it rightly questioned the numerous claims "that Canadian firms have been as aggressive acquisitors of foreign firms as foreign firms have been of Canadian companies."[23]

Sitting on my desk is a pile of misleading press clippings, too many to detail, but I'll mention two examples. In April 2007, *Report on Business* told its readers that, in 2006, 123 companies in Canada were taken over by foreigners. Interesting, but the correct number of takeovers was 363, quite a difference. On Canada Day, 2007, Gordon Nixon and Roger Martin told *Globe and Mail* readers that there were only 455 foreign takeovers of Canadian companies from 2001 to 2006. In fact, there were 1,441 takeovers. Again, quite a difference.[24] One would think that people like Nixon and Martin would try to get their numbers right.

Industry Minister Jim Prentice repeats the ridiculous Mulroney slogan "Canada is open for business but not for sale," a position which the *Globe and Mail* described as a "balanced approach."[25] Prentice repeats the nonsense, across the country, that Canadian companies have been buying more foreign companies than foreigners are buying Canadian companies. When challenged by me, Prentice's Ottawa officials claimed that in the period from 1996 to June 2007, Canadian companies acquired 3,898 foreign companies, while foreign companies acquired only 1,540 Canadian companies. Not one single journalist or member of Parliament was curious enough to pick up the telephone and check Prentice's numbers with Investment Canada.[26] Had they done so, they would have found that, yet again, the Harper government was intentionally misleading Canadians on the issue. The correct number of takeovers of companies in Canada was not 1,540 but 6,355.

As a further contribution to the correction of corporate and political BS, let me note that in the 10 years after the Foreign Investment Review Agency was established, real GDP increased at an average annual rate of 3.54 percent. In the 10 years after Brian Mulroney dumped the Foreign Investment Review Agency for Investment Canada, real average annual GDP increased at an average rate of only 2.52 percent.

So much for all the nonsense about how controls on foreign direct investment damaged the Canadian economy.

I'll leave the last words to former Deputy Prime Minister Paul Hellyer, former Quebec Industry Minister Rodrigue Tremblay, and financier Stephen Jarislowsky. For Hellyer, "I tried to find a nicer word than treasonous to describe what is going on, but to no avail. It is time to call a spade a spade and let our politicians – and the 'experts' who advise them – know in no uncertain terms that we hold them accountable for their unforgivable negligence."[27]

For Tremblay, "Canada as a whole is in danger of losing control of its most important economic levers. . . . The slogan 'Masters in our own house' seems very remote these days. . . . What is happening surely appears to be contrary to Canada's best interests."[28]

For Jarislowsky, "What's happening is 'economic suicide.'"[29]

FOREIGN TAKEOVERS

"It's a disaster for Canada. Anything and everything is
for sale. We'll run out of companies to invest in."[1]

I
n early 2006, business columnist Andrew Willis of the *Globe and
Mail* wrote:

> Virtually every company in this country without a controlling
> shareholder seems to be on the auction block these days.
>
> The unanswered question is just how the people of
> Canada, and their elected officials in Ottawa, feel about
> selling off the biggest businesses in the land.
>
> Expect takeovers to continue at a red-hot pace.

And, of course, Willis was right. It's been hard to open a daily newspaper
without a story about yet another foreign takeover. As to how Canadians
feel, the evidence is mixed. As mentioned earlier, in all public opinion
polls, for many years, Canadians overwhelmingly have said we already
have too much foreign ownership and foreign control and don't need
more. Yet there has been relatively little in the way of a public outcry as
more and more corporate icons fall.[2]

How does Ottawa feel? Don't expect much change. The chances of
the Harper government doing anything of real importance about
growing foreign ownership and control are at best remote. When asked
about it in June 2007, the prime minister said he had "no intention of

intervening." Perhaps Lynton (Red) Wilson and his policy review panel appointed in the summer of 2007 to study the problem will have enough influence with Harper to prove me wrong. Perhaps. Certainly there's a chance that strong public opinion might make a difference. Yet if the Liberals regain power in Ottawa, based on their record in office during the Chrétien and Martin years there's little evidence that they will be much better.

According to Stéphane Dion, "Canada is not for sale [sound familiar?]. I believe domestic ownership does matter. . . . I promise to . . . protect our economic sovereignty."[3] Very nice. But it would have been nice to have heard from Dion during all the Chrétien and Martin years when so many companies were vanishing from Canadian control. Dion later reassured the business community that he was "no protectionist."

As David Olive wrote in 2006 in the *Toronto Star,*

> In the past two years, more than a dozen of Canada's largest corporations with total assets of more than $57 billion have been swallowed by foreign predators. And the response has been . . . well, there really hasn't been a response.
>
> There have been no calls in Parliament, any legislature, the media, or the halls of academe for a royal commission on the consequences for our economic sovereignty from this unprecedented yard sale of Canadian industrial assets.

So, we have

> the outsourcing of decision-making to Switzerland and California, the sapping of the Toronto Stock Exchange as listings disappear: the decline in business for ancillary providers such as legal, accounting and underwriting firms and the implications for philanthropy as CEOs focus on the communities in which they live . . . [leaving] too many absentee owners of glorified shipping depots.

Meanwhile, "Our G-7 partners France, Germany and Italy regularly thwart attempted outside takeovers deemed to be of strategic national importance."[4]

The appalling hypocrisy of most of Canada's business leaders regarding the foreign takeover of the Canadian economy is matched only by the gullibility of some of our business journalists. For example, a *Financial Post* writer quotes a CEO as suggesting that even more tax cuts "might boost profits and create 'a powerful incentive' for companies to remain independent." Nowhere in the article does the writer mention the successive years of record corporate profits in Canada, or the fact that yet lower taxes would be even more likely to encourage foreign takeovers than the substantial decreases in taxes have done already.

Nor does the writer seem to understand that, in the first half of this decade, there was already a record number of foreign takeovers, to be easily topped in 2006, even with no further tax reductions. And 2007 seems to have been yet another disaster year for Canadian economic sovereignty.

But true to form, as the *Financial Post* has told us, "Canadian business leaders want the federal government to cut business taxes to keep Canadian companies from selling operations to foreigners a new survey done for the *Financial Post* reveals."[5]

Yet another poll, reported this time in the book *What Canadians Think,* by Darrell Bricker and John Wright of Ipsos Reid, some 70 percent of Canadians agreed that the "Federal Government should have the power to stop an American company from purchasing a Canadian company," and six in ten said that they were "angry that the Federal Government is not doing more to stop U.S. and other foreign ownership from buying Canadian-owned companies."

Unfortunately, the Mulroney government paid zero attention to many similar polls, and neither did the Chrétien or Martin government. As for the Harper government, you've already read what I think is happening and what I think will happen in the future. In October 2007, Stephen Harper, true to form, said that the fear of foreign takeovers was unwarranted. Welcome to the branch-plant colony of Canada. One week after

Harper dismissed any concerns about growing foreign ownership and control, a new study confirmed that Canadians were "losing corporate control to outside investors."[6]

Among the Canadian companies taken over since the long depressing list I put together in *The Vanishing Country* in 2002 are the Hudson's Bay Company, Inco, Alcan, Stelco, Placer Dome, Falconbridge, mountain resort operator Intrawest Corp., Newbridge Networks, Dofasco, Geac Computer Corp., Leitch Technology Corp., Creo, Vincor International, Sleeman Breweries, Gale Force Energy, EnCana's holdings in the Mackenzie Delta and Canadian Arctic, AIM PowerGen Corp., BCE's Telesat Holding, the lingerie retailer La Senza Corp., Calgary's Centurion Energy International, the Toronto-based luxury hotel operator Four Seasons Hotels & Resorts, Ontario auto parts maker Meridian Technologies, THINKFilm, Harris Steel Group, Fanny Bay Oysters, Petro-Kazakhstan, Domtar, ATI Technologies, Ipso, Liton Ore International Mining, Prudential Steel, Co-Steel, Arcelor, Algoma Steel, Voxcom Income Fund, UE Waterheater Income Fund, Nelson Resources, LionOre Mining, Standard Aero Holdings, Vancouver Wharves, TIR Systems, Halterm Income Fund, Lakeport Brewing Income Fund, Norcast Income Fund, Entertainment One Income Fund, Great Lakes Carbon Income Fund, KCP Income Fund, Gateway Casinos, Calpine Power Income Fund, ATI Technologies, BioChem Pharma, ID Biomedical Corp., the film distributor MPD, jam and sauce maker E.D. Smith & Sons, the refrigeration company Versacold, the Winnipeg Commodity Exchange, the high-tech icon Alias Systems, Limocar, Air Canada Technical Services, Steeplejack Industrial Group, Aspreva Pharmaceuticals, Western Oil Sands, Prime West Energy Trust, Vitibev Farms, Westwind Partners, Cognos, and Miramar Mining, with potentially the largest gold mining area in Canadian history, Axcan Pharma and MacDonald Dettwiler and Associates.

After the U.S. giants Cargill and Tyson arrived in Canada, most of our meat packers disappeared. Canadian agrologist Wendy Holm says that the two U.S. firms controlled 65 percent of feed cattle slaughter in Canada, and with the 2005 takeover of Better Beef in Guelph, Ontario, they now control over 80 percent. Gone, too, are most of our big mining companies, all of our steel companies, and even our big breweries.

Yet Finance Minister Jim Flaherty proclaimed in 2007 that in the future he wants much more of the same. And according to the *Globe and Mail*, "Industry Minister Maxime Bernier is laying out a welcome mat for foreign companies to acquire Canada's telecommunications companies."[7]

The *Globe*'s Eric Reguly asks,

> Why is Canada apparently happy to have industries owned by companies and people who don't live here?
>
> Canada is one of the few western countries to have never stopped a foreign takeover. The Americans stop them all the time.[8]

Dominic D'Alessandro, chief executive officer of Manulife Financial Corp., is one of only a few senior Canadian business executives who have spoken out about the fire sale of our country.

> We just assume that anybody should be allowed to buy our assets, buy our companies. . . .
>
> There are even professors now preparing papers saying we haven't really lost anything, that a change of ownership doesn't matter. . . . I think it matters a lot.
>
> I don't think we should be embarrassed as a population to say that there are certain sectors that we should reserve for ourselves.

And, soon after:

> I sometimes worry that we may all wake up one day and find that as a nation, we have lost control of our affairs.
>
> I find it particularly bothersome that so many of our natural resource companies are now owned elsewhere.

What if we were to consider adopting ownership restrictions for certain sensitive sections of our economy that would be similar to those that now apply to our financial institutions?[9]

In 2006, before another buyout binge in 2007, almost one fifth of Canadian stock market capitalization had been taken over by foreign corporations. Yet the very unhappy law firms, insurance firms, telecommunications firms, etc., etc., that are affected by all this behave as if their mouths were bound with tape.

Early in 2007, I became increasingly dismayed by how few of our corporate elite were speaking out about what was happening. I called four very prominent Torontonians, one at the Rotman School of Management at the University of Toronto, another a much-admired and very prominent Canadian executive, another a well-known business writer, and the fourth one of Canada's most famous entrepreneurs, and I asked them if there was *anyone* speaking out about the foreign takeover of our country. The response was both sad and significant. Here is the e-mail I sent to Royal Bank CEO Gordon Nixon after talking to all four:

> Mr. Nixon:
>
> In recent talks with Red Wilson, Roger Martin, John Evans and Dominic D'Alessandro I asked them why there are not more people from corporate Canada speaking out about the rapidly increasing foreign ownership and foreign control in Canada. (Aside from Dominic that is.) I asked them if they could name anyone who was. All four named you.
>
> I would welcome having a copy of your remarks on this subject.
>
> Mel Hurtig, O.C.[10]

Nixon responded by sending copies of some of his recent remarks, which included this: "We have not only seen the disappearance of major

Canadian household names, but the loss of Canadian presence in industries where we have long had traditional strengths. Their loss does not bode well for Canada's future performance."

This response was gratifying, but it also made me curious. I wrote back to Mr. Nixon, congratulating him for his comments, but also asking, given his expressed concerns and the congratulations he has received for making them, several questions.

1. Are you now prepared to see to it that the Royal Bank stops its long-term practice of helping to fund the foreign takeover of Canadian companies?
2. Since the Royal Bank was probably the most enthusiastic and significant single corporate supporter of both the FTA and NAFTA, were you not aware of the investment clauses in both agreements which make it much more difficult to block American takeovers of Canadian companies?
3. You are now the head of the Canadian Council of Chief Executives. When you accepted this premier big-business position, did you not know that both the CCCE and its predecessor the BCNI have been and are the most rabid and enthusiastic supporters of unrestricted foreign direct investment?
4. How can you reconcile your public concerns about increasing levels of foreign ownership and foreign control with these three questions?

Gordon Nixon has not replied to my questions. But, remarkably, in an interview he has said we should not go as far as actually putting restrictions on foreign ownership.[11] How bizarre! And this is the man corporate Canada points to as a protector of Canadian ownership and Canadian control. But then again, since he's the head of the CCCE, what else could we expect? In November 2007, the *Financial Post* named Gordon Nixon "Canada's Outstanding CEO of the Year."

I suspect that after this book is published we may not be hearing much on the subject from Mr. Nixon. But if any of you attend any of his speeches, you may wish to ask him some of the above questions during the Q & A session, particularly the first question.

For well-known and widely respected petroleum industry executive Dick Haskayne, the Canadian culprits are "ego-driven executives and directors, fee-grabbing investment bankers, loose-pursed lenders and the governments whose policies foster such transactions."[12]

For Peter Munk of Barrick Gold Corp., the deficiencies of our business leaders are obvious: "Now look what they've got. It requires balls, it requires guts, it requires vision. And those are not qualities that come to our senior corporate managers."[13]

Should Canadians be concerned about the current surge of takeovers? Even the traditionally continentalist editorial page of the *Globe and Mail,* which for decades has strongly decried any barriers to more foreign direct investment, has called the rate of takeovers "alarming" and in a truly astonishing out-of-character editorial actually criticized the Harper government for having

> no coherent plan to safeguard Canada's strategic industries. Indeed, it has yet to signal whether it considers any industries vital to this country's future and hence essential to keep under Canadian control.
>
> Consolidation in the resource sector means job losses, and the reduction of head offices means less demand for talent in a wide range of financial, legal and technical services.
>
> The real decisions will inevitably be made elsewhere by managers who own no allegiance to Canada or its strategic interests. The promise that head offices will stay in Canada is as hollow as the offices themselves will soon be.[14]

Wow! Amazing! What a sharp contrast to the *Globe*'s long-standing position. And how very timely and welcome.

But Peter C. Newman rightly remains pessimistic: "There's no popular outcry, there's no popular whimper. When those leading companies disappear, we lose our membership card in the global economy. We're becoming a kind of Manchuria, supplying raw materials to the more mature world."[15]

Eric Reguly says, "Canada is a G8 country that acts like a colony, circa 1900, waiting to be plundered by the Americans, the Europeans, even the Latin Americans." And why is this happening? According to Reguly, "Feckless CEOs are the main problem."[16]

The following letter appeared in the *Globe and Mail,* May 5, 2007:

Memo to Dominic D'Alessandro, CEO of Manulife (CEO Urges Action on Takeover Frenzy – Report on Business, May 4):

Your concerns about Canadian ownership reflect a deeper question, namely the level of ability and leadership (or lack thereof) among the ranks of our business leaders.

The recent actions of a number of CEOs seem to suggest it is easier for them to sell and run (with pensions in hand), rather than lead. "If they were on a battlefield," as my grandfather, a First World War veteran, would have said, "they'd all be shot!"

Leo J. Deveau, Wolfville, N.S.

CANADIAN INVESTMENT ABROAD

C anadians were solemnly promised that the Canada-U.S. free-trade agreement (FTA) would result in massive increases in new foreign firms locating in Canada, since our country now had "guaranteed access" to the giant U.S. market. Also, much new domestic Canadian investment would be stimulated to take advantage of this "privileged" position.

It didn't exactly work out that way. Instead, big business in Canada, led by our "Big Five" banks and our other financial institutions, began pouring huge quantities of money out of the country, mostly into the United States, while over $53.9-billion went into tax havens in Barbados and Bermuda. By the end of 2006, over 44 percent of Canadian direct investment abroad (CDIA) was by our finance and insurance sector.

It's very revealing to compare CDIA during the first 13 years (from 1994 to 2006) of the North American free-trade agreement (NAFTA) with the preceding 13 years (1981 to 1993). In the NAFTA years, CDIA totalled an enormous $499.9-billion. This compares with an outflow of only $75.6-billion during the previous 13 years.

So much for the con artists who promised that NAFTA would encourage Canadian corporations to invest here. It was the biggest, most enthusiastic NAFTA promoters (think of our Big Five banks again) who began pipelining money out of the country as fast as they could, beginning in 1994. Prior to NAFTA, the record year for CDIA was in 1987, when outflow amounted to $9.44-billion. During the NAFTA years, the outflow has exceeded $50-billion in five different years.

During the years 2002 to 2004, only four countries had a greater outflow of direct investment than Canada. In these years, Canada's outflow was greater than the combined foreign direct investments for all the following countries *put together:* Denmark, New Zealand, the Slovak Republic, the Czech Republic, Poland, Greece, Turkey, South Africa, Hungary, Iceland, Finland, Germany, India, Mexico, Norway, Korea, Brazil, Portugal, and Austria. And it was greater than 10 other countries, namely Russia, Ireland, Australia, Sweden, Italy, Switzerland, China, the Netherlands, Belgium, and Japan.

Between the beginning of the FTA and the end of 2006, Canadian direct investment in the United States increased from $56.6-billion to $223.6-billion. At the end of 2006, it amounted to $59-billion in the United Kingdom, $24.7-billion in Ireland, $16.9-billion in France, and just over $12-billion in the Netherlands. Investment in the United States accounted for 43 percent of the total.

Yet it bears repeating that, contrary to so much misinformation on the subject, Canada's direct investment in the United States has never been greater than U.S. ownership and control of assets in Canada, and it has always been a very much smaller fraction of GDP.

According to the Conference Board of Canada, more and more Canadian businesses seeking to expand into the U.S. market would rather build new facilities there than increase existing operations in Canada. This despite lower wage, land, utilities, and resource costs in Canada, and corporate taxes that have for some time been generally lower in Canada. And, of course, many of these very companies were exactly the ones so ardently promoting the FTA and NAFTA and the two agreements' certain lavish investment benefits for Canada.

Until the first year of the FTA, the greatest amount of Canadian direct investment abroad in any one year was under $10-billion. In the years since 2000, it has ranged from a low of $25.6-billion to a high of $66.4-billion in a single year.

While in the free-trade years Canadian direct investment has been pouring out of Canada, the same is also true, in gigantic amounts, for portfolio investment. To the end of 2005, the amount invested outside of Canada came to $368.3-billion.

In the 2005 budget, the federal government allowed RRSP money to be invested outside of Canada. At the same time, Ottawa increased RRSP contribution limits to $22,000 by 2010. There is nothing wrong with allowing Canadians to invest outside of Canada, but it's totally absurd to give them tax breaks (by means of RRSPs) to do so.

Meanwhile, our Canadian banks, with their record-breaking profits, are pumping money out of Canada into tax havens like Barbados, Bermuda, the Cayman Islands, the Bahamas, and Ireland. Between 1990 and 2003, Canadian assets in tax havens shot up from $11-billion to $88-billion. Canadian banks accounted for $72-billion of that $88-billion. By the end of 2006, total Canadian direct investment in tax havens was close to $100-billion.[1]

In a surprise move in February 2005, the Liberal federal government removed limits on foreign content in pension plan investments, restrictions that had been in place for over 34 years. Some leading investment executives said they were shocked by the unexpected government decision, and financial analysts predicted that a great deal of investment now managed in Canada would be managed in the U.S in the future, making more than a few major investment firms in Canada "an endangered species." (It's important to note that few Canadians can take full advantage of the RRSP limits. In Statistics Canada's most recent report, for 2005, when limits were $16,500, only 31 percent of RRSP investors were able to do so in that year.)

With the elimination of restrictions on foreign investment by pension funds and RRSPs, many large institutional investors have increased – or were planning to greatly increase – their investments outside of Canada. In June 2005, it was revealed that a "secret" Finance Department study estimated that, in the following two years, almost $70-billion would flow out of Canada as a result of the Martin government's decision to lift the long-standing 30 percent cap on the foreign content of pension funds.

Finance was way out. In 2006 alone, Canadians invested a record $78.3-billion in foreign securities during the first full year when there was no limit on the foreign content of RRSPs and pension plans. So it wasn't only Canadian direct investment abroad that exploded when NAFTA came into effect. In the 13 years before NAFTA, Canadian portfolio investment abroad in stocks amounted to $51.8-billion. In the first 13 NAFTA years, it increased by over $257-billion, while Canadian investment in foreign bonds during the same years increased from $13.76-billion to $127.1-billion. Canadian investment in foreign bonds was a record $43.6-billion in the one year of 2006, far greater than in all the 13 pre-NAFTA years put together.[2]

In sum, we have more and more tens of billions of dollars hemorrhaging out of the country, mostly from our banks and other big corporations, plus the large pension funds, while more and more non-Canadians take over the ownership and control of our country, often using financing they have obtained from the very same banks and financial institutions.

Sheer suicidal lunacy!

And as for the ridiculous claims of the Canadian Chamber of Commerce and the Conference Board of Canada, among others, that we must not tamper with foreign takeovers of business in Canada because we need the money, one can only juxtapose such claims with the news of the enormous multi-billion dollar losses by our Canadian banks in U.S. subprime mortgage related securities.

Lastly on the subject of Canadian direct investment abroad, it's very important to note that a great deal of the so-called "Canadian" investment is not Canadian at all. Foreign firms based in Canada borrow money from Canadian banks and then deduct the interest charges from their profits in Canada to reduce their taxes here. Then they invest the money abroad, often in tax havens. This both inflates "Canadian" direct investment abroad and sharply reduces taxes payable in Canada. Keep this in mind when you repeatedly hear our compradors claim that we shouldn't be concerned about foreign ownership and control when our "Canadian" investment abroad is so high.

THE FREE TRADE AGREEMENT

THE MOST COLOSSAL CON JOB IN CANADIAN HISTORY

According to recent editorials in the *Globe and Mail,* Brian Mulroney's "free trade agreement with the United States laid the groundwork for economic prosperity"[1] and prepared the way for "immense economic benefits."[2]

This has been the *Globe*'s editorial line for as long as I can remember. It's also been the unquestioning, jubilant refrain from the Canadian Council of Chief Executives (the godfathers of the FTA when they were the BCNI), the Canadian Chamber of Commerce, the Conference Board of Canada, Canadian Manufacturers & Exporters, and of course the C.D. Howe and Fraser Institutes (not to mention the numerous Milton Friedman economists in our universities).

Aside from their blind faith in the "obvious" success of the Canada-U.S. free-trade agreement (FTA), all the above organizations have two important things in common. They are all mostly funded by the same big right-wing corporations (and, as we have seen, a large percentage of these are American). Second, none of them is apparently willing to believe the extensive, detailed documentation of the actual economic results of the FTA produced by Statistics Canada.

No corporation pushed Mulroney's free-trade agreement with the United States harder than the Royal Bank of Canada (which, of course, now promotes itself as "RBC" so it can more easily invest even more of its, and our, money in the United States).

Is it any wonder that, in the face of the increasing and overwhelming evidence that the FTA failed to produce the results Canadian big business and our negotiators had promised, the bank produced its own laudatory version of the agreement's results in a widely publicized November 2006 report. "Free-Trade Fears Unfounded, RBC Says" was a headline in the *National Post*. Fears and criticisms are "myths" of the past. And, please note, "Exports have soared and foreign direct investment in Canada has risen substantially."[3] (No mention of how many billions of dollars the Royal Bank ships out of Canada or how much they lend to foreigners to help them take over Canadian companies.)

During the past decade, a consistent picture has been presented by big business, by almost all of our Liberal and Conservative politicians, and in most of our business pages. The FTA has been an enormous success, producing "massive benefits,"[4] a great, unquestioned triumph. Given the large increase in Canadian exports, how could anyone possibly think otherwise? To say that the media in Canada have completely and miserably failed and misled Canadians in respect to the economic results of the FTA is no exaggeration.

It's too bad that apparently none of these people knows how to operate a simple calculator. And it's too bad that none of them is familiar with the huge amount of contrary evidence in numerous publications such as Statistics Canada's *National Income and Expenditure Accounts,* or the excellent Statistics Canada annual, *Canadian Economic Observer: Historical Statistical Supplement.*

Where to begin? Let's start with GDP per capita. In the 17 years before the FTA, 1972 to 1988, GDP per capita increased by $9,365. During the first 17 years of the FTA, 1989 to 2005, the per-capita increase fell to only $8,354.[5] In the 17 years before the FTA, Canada's overall GDP increased at an annual rate of over 3.66 percent. During the first 17 years of the FTA, it increased at an annual rate of only 2.63 percent. Quite a difference!

In the years before the FTA, Canada's GDP per capita was as high as 91.5 percent of the U.S. rate. By 2006, it had fallen all the way down to 83.8 percent. During the 1960s, Canada's annual GDP percentage increase was greater than that in the United States in eight of the 10 years. In the 1970s, it was greater than the United States in six of the 10 years. However, since the FTA, during the 18 years ending in 2006, the U.S. GDP increase was greater than Canada's during 12 of the 18 years.

There have been numerous claims that the real objective for Canadian FTA negotiators was to come away with an agreement that would raise the rates of productivity in Canada. In fact, Canadians were frequently promised this result during the free trade debates of the mid- to late-1980s.

Let's see how we made out. In the 17 years before the FTA, output in the business sector in Canada increased at an average annual rate of 3.31 percent. However, from 1989 to 2005, it increased by only 2.84 percent a year.[6] What about output per hour in the business sector in Canada compared to the United States? Before the FTA, it was as high as 90 percent. By 2006, we had fallen all the way down to less than 74 percent.

And how have we made out in terms of the annual rate of change in productivity in the total Canadian economy? In the 17 pre-FTA years, it increased by an average of 3.63 percent a year. In the first 17 FTA years, the annual increase fell to only 2.65 percent. What about labour productivity levels in Canada as a percentage of those in the United States? In 1988, we were at about 90.5 percent of the U.S. level in GDP per worker. By 2006, we were down to under 84 percent. In GDP per hour, we were over 88 percent of the U.S. level in 1988, but down to only 82.4 percent in 2006.

What about the creation of new jobs? In the 17 pre-free-trade years, we created 4.43 million jobs in Canada. In the first 17 years of the FTA, we created only 3.41 million jobs. In the 17 years before the FTA, employment in Canada grew at an annual average rate of 1.76 percent. From 1989 to 2006, the annual growth rate fell to 1.41 percent. Once again, a substantial difference.

In every single year from 1989 to 2005, Canada's unemployment rate was higher than the OECD average.[7]

While real exports of goods and services have increased during the free-trade years, employment growth has nowhere near matched exports in any comparison you can think of. Moreover, job growth in Canada has been much greater in domestically oriented industries than in export-oriented sectors of the economy. In fact, one study prepared for Industry Canada showed that increased trade had actually produced a net loss of jobs, because increased imports displaced more jobs than increased exports generated.

Now let's turn our attention to how workers in Canada have made out since the FTA came into effect in 1989. In the 10 years before the FTA, compensation per employee in the Canadian business sector increased by an annual average of 7.4 percent, slightly ahead of the total OECD average of 7.2 percent, and also ahead of the U.S. average of 6.1 percent.

From 1989 to 2005, however, the average compensation increase per employee in Canada over the 17-year period was only 3.25 percent, well under half of the average for the decade before free trade, and now below the OECD average of 3.85 percent, and also below the U.S. average of 3.95 percent. Measured another way, negotiated wage settlements in the first 17 years of the FTA produced an average annual increase of 2.34 percent. During the same years, the consumer price increase averaged 2.43 percent.

During the 1980s, the average annual increase in employee wages in Canada was 6.34 percent. From 1990 to 2006, this fell all the way down to 3.27 percent. Many opponents of the FTA and NAFTA have contended that both agreements were primarily meant to drive up corporate profits by suppressing wages, with threats that jobs would be transferred to Mexico or to low-wage U.S. states. Certainly, labour income results support this theory.

While it's true that many Canadian are much better off today than in 1989, it's also true, as we've seen earlier in this book, that many in the middle and lower income brackets lost ground. Let's see how personal income in Canada fared during the first 17 years of free trade and compare the average annual percentage change with the 17 years before the FTA.

In the 17 years before free trade, the average annual increase in personal income was 11.7 percent. During the first 17 years of free trade, the annual average plummeted to 4.16 percent, once again a huge difference. In 1988, the year before the FTA came into effect, personal income in Canada was just over 87 percent of the U.S. per-capita rate. By 2006, it was down to only 80.4 percent of the U.S. rate.

A similarly dramatic impact relating to personal income can be seen by measuring the personal savings rate. In the 17 years before free trade, it averaged 14.4 percent. During the first 17 years of free trade, it dropped to an average of 7.28 percent. For the 10 years from 1996 to 2005, it was 4.1 percent. By 2006, it was all the way down to 1.6 percent. Even with an increased population, the total of all personal savings in Canada in 2005 was only 15 percent of what it was in the early 1990s.

In 1988, the year before the FTA came into effect, the household savings rate in Canada was 12.3 percent of disposable household income. In the 17 years since 1988, the average annual savings rate fell all the way down to 7.1 percent; by 2004, it was 1.4 percent; and by 2005 it was in a negative position of −0.2 percent. The savings rate has been above 5 percent only once during the past nine years, and the OECD has predicted a savings rate in Canada of only 1 percent in 2007, far below the rates for countries such as Austria, France, Germany, Italy, the Netherlands, Norway, Sweden, Switzerland, Belgium, Portugal, Spain, and the United Kingdom.

Then there are those who have pointed to our high non-farm capacity utilization rate of 86.1 percent in 2005 as proof that the free-trade agreements have been a big success. But in 1988, the year before the FTA kicked in, the capacity utilization rate in Canada was 86.6 percent.

And the income of Canadian families? Since the FTA came into effect, the average annual market income of the lowest 20 percent of Canadian families dropped by almost $400; the next 20 percent lost an annual average of $2,900; the next 20 percent lost $300; while the second highest 20 percent of families had an increase of $300 a year, and the top 20 percent had annual average increases of $19,000.

Put another way, Statistics Canada reports that since the start of the FTA, "Average family market income among the 10 percent of families with the highest incomes rose by 22 percent from 1989 to 2004. Meanwhile, among the 10 percent of families with the lowest incomes, it fell by 11 percent."[8]

Another study, this one for the province of Ontario, showed that since 1989, 90 percent of Ontario's families with children under the age of 18 saw a drop in their real incomes, ranging between $5,000 a year in constant dollars to $9,000, depending on the income level.[9] Try that again. Since 1989, 90 percent of Ontario families with children have lost ground!

Another useful way to gauge the impact of free trade on our economy is by measuring the average annual increase in millions of dollars in building permits. In the 17 years before the FTA, these annual increases averaged some 12.2 percent. In the first 17 years of free trade, they averaged only 3.87 percent, once again a big drop.

Since the FTA came into effect, everyone who promoted the agreement has trumpeted its success by pointing to the large increase in exports, mostly to the United States (78.9 percent of our total exports in 2006). Fair enough. The same people, however, never mention the very large increased foreign content of these exports, nor the fact that much of the increase in exports has come from the export of our rapidly depleting reserves of natural gas and conventional oil. Nor do they ever mention the Industry Canada study showing that well over 90 percent of our export increases were the result of energy exports, plus the value of the Canadian dollar, which had been so low for so many years.

Beyond that, it's worth looking at the actual annual percentage increases of exports, comparing them with the pre-free-trade years. In the first 17 years of free trade, Canada's merchandise exports increased by an annual average of 7.2 percent. However – surprise, surprise – in the 17 years before the FTA, merchandise exports had increased by an annual average of 13.4 percent.

Of course, we are constantly told about the large increase in exports since the FTA. But, again, what about imports? In 1988, imports of goods and services were $159.1-billion. By 2006, they had increased to $487.7-billion. Well, surely, with our huge increases in exports to the United States under free trade our share of the U.S. market must have increased.

Not so, it declined. In 1988, the year before the FTA came into effect, Canada's share of total U.S. imports was at 14.88 percent. By 2006, it was down to 12.6 percent.[10] Once again, exactly the reverse of what we were promised.

It's not only a large percentage of Canadians who now doubt the benefits of free trade. In December 2005, a survey commissioned by the transatlantic German Marshall Fund of the United States found that the majority of those surveyed in the larger European countries and in the United States as well felt that free trade reduced local employment and benefited mostly big multinational corporations. In late December 2007, the CCPA released a study showing that since the FTA was signed, a sample of CCCE companies cut their workforce by just under 20 percent (over 118,000) while their revenues grew by 127 percent.

As I and many others have pointed out, Brian Mulroney won a majority government in the 1988 "free trade election" even though a majority of Canadians voted for parties opposed to the agreement, and the Conservatives won a majority of the seats in only two provinces, Quebec and Alberta, "the two provinces least favourable to the Canadian state," in the words of economist Erin M.K. Weir.

Weir repeats a number of points some of the rest of us have made:

- we gave away the farm at a time when American tariffs only amounted to less than 1 percent of the total value of Canadian exports to the United States;
- we completely failed to gain our major objective of exemption from U.S. trade-remedy laws;
- the U.S. implementation legislation actually made it easier for U.S. companies to initiate countervail against Canadian products than against other foreign products;
- Canada did so poorly because it surrendered its bargaining chips before negotiations began;
- by any measure, Canada was out-negotiated, and extracted almost no meaningful concession from the United States through free trade.

If any of you want to send Brian Mulroney, Thomas d'Aquino, Stanley Hartt, Pat Carney, Derek Burney, Michael Wilson, Simon Reisman, Allan Gotlieb, Wendy Dobson, the Canadian Chamber of Commerce, the Conference Board of Canada, the C.D. Howe Institute, or any other FTA promoter a calculator for Christmas, please let me know and I will be happy to send you their addresses.

NAFTA

"Free trade is much less than it seems;
the NAFTA emperor has no clothes."

"The American refusal to comply (with the softwood
lumber decision) tears the heart out of NAFTA."

L et's start off with another *Globe and Mail* editorial, this one from
2006: "The benefits of the Canada-Mexico-U.S. deal have been
huge.... NAFTA is responsible for much of Canada's prosperity."[1]
Later in the year, another *Globe* editorial told readers that "after all, the
North American free trade agreement has brought prosperity to all three
partners, including the U.S."[2]

In the previous chapter, we saw the "huge" supposed benefits of free
trade for Canada as a result of the FTA. If anything, the results of the
North American free-trade agreement (NAFTA) are worse and, because
of Chapters 6 and 11 in the agreement, are certain to be even more
harmful in the future.

About the same time as the second *Globe* editorial appeared, a major
new public opinion poll in the United States confirmed that most
Americans now opposed free trade. In Mexico, every year, there has been
growing disenchantment and growing resentment. In Canada, poll after
poll has shown that Canadians believe Canada has been the loser in
NAFTA and the United States the big winner.

Despite this, Thomas d'Aquino of the Canadian Council of Chief
Executives, appearing before a Canadian Senate committee, said, "The
level of support for NAFTA is highest here among the three countries."[3]
Not one of the senators was informed enough to respond, "But that's not
saying very much, is it?"

Today in Mexico, some three-quarters of the population live in poverty. Between the time NAFTA came into effect, on January 1, 1994, and 10 years later, real wages for Mexican workers fell by over 14 percent. NAFTA increased employment in the very low-wage *maquiladora* regions, while the important agricultural sector saw a steady loss of employment. Overall productivity growth has been virtually non-existent, and labour productivity is only about one-third of the OECD average.

For Canada, it's true that Canadian exports to Mexico have increased, but our imports from Mexico increased much more. Whereas our trade deficit with Mexico was about $3-billion before NAFTA, it's now headed for $12-billion. The previous chapter contained a devastating analysis of the impact of the very poorly negotiated FTA. Shortly, we'll hear some remarkable mea culpa comments from mostly the very same people who were involved in doing an even poorer job regarding NAFTA.

It's difficult to understand how supposedly intelligent people concerned with the future welfare of their country could ever have signed on to such truly awful agreements as the FTA and NAFTA. The sight of them doing their embarrassing explanations on the front page of the *Globe and Mail* in the summer of 2005, after the softwood lumber surrender, would have been amusing if the whole fiasco had not been so tragic and so dangerously precedent-setting.

The NAFTA agreement contained provisions that no other developed country would have accepted. Chapter 6 is incredible. Somehow our negotiators agreed to a host of ridiculous binding provisions that are completely incomprehensible for anyone concerned with Canada's future. Here, I'll touch on only two of the most egregious.

Even if, in our large, cold, northern country, we Canadians begin to run short of oil and natural gas, under NAFTA we must continue to supply the United States with a guaranteed pro-rata share of our production. If we want to cut back our exports to the United States, we must cut back our own consumption in an equivalent manner. And – equally incredible, to the point of being totally absurd – we cannot charge the Americans a penny more for our oil and gas than the price Canadians pay. At this writing, Petro-Canada is running short of Canadian natural gas and is trying to secure a steady supply from Russia. It truly boggles the mind.

Then there's the by-now-notorious Chapter 11, whereby U.S. firms or individuals are allowed to sue Canadian governments for legislation which the Americans believe may decrease their profits. So, if in Canada the federal or a provincial government takes action to protect the environment, or broaden medicare, or bring in public auto insurance, or enact any number of other measures, the government may be successfully sued by U.S. corporations. In September 2007, Exxon Mobil and Murphy Oil served notice that they plan to use NAFTA to sue Canada because Newfoundland and Labrador is forcing them to spend more energy research dollars in the province, and an American couple have announced plans to sue because of the Harper government's actions regarding energy trusts.

Talk about abandoning sovereignty! Talk about voluntarily opting for a colonial status.

In the investment chapter in this book I touched upon another serious negative that NAFTA brought us. This is what *The Economist* had to say about the agreement: "The default option for multinationals (including Canadian-controlled multinationals) chasing the mighty American market is to invest in it directly. Indeed, NAFTA has strengthened this preference. American firms that used to open plants in Canada to sell into its protected market no longer need to."[4] (Reader, if you don't mind, please read that quote again, to make sure it sinks in.)

Moreover, because of NAFTA, many of the normal nation-building tools that could and should be used to develop our country, tools insisted upon by many other countries, are no longer available to us. To take just one example, one out of dozens: Obliging foreign corporations to do some of their purchasing locally is no longer possible.

The respected former Ottawa trade negotiator Mel Clark has pointed out that for 40 years, under General Agreement on Tariffs and Trade/World Trade Organization rules, the United States could not attack Canadian lumber exports the way they have under the FTA and NAFTA. During those years, Canada won most trade disputes with the United

States. But now, it is essentially American law that applies to all our exports to the United States. And as we have learned, the dispute settlement panels we reluctantly agreed to when the United States refused to abandon its countervailing and anti-dumping penalties have proved to be worthless when the United States refuses to obey their rulings.

Moreover, under the terms of the two poorly negotiated agreements, the United States can change and toughen its trade laws in relation to Canada however and whenever it wishes to do so. And as we have seen, they can continue to ignore unanimous NAFTA rulings whenever they wish. There's not much doubt about the bottom line regarding the U.S. attitude towards trade law. The United States believes in free trade, as long as the benefits accrue to Americans.

The softwood lumber fiasco made it clear that Americans regard the NAFTA dispute settlement process as somewhere between largely unimportant and completely irrelevant. The clear set of rules Canadians wanted when we negotiated the agreement, and the exemptions to U.S. countervailing and anti-dumping laws, were not to be had. (It's very difficult to understand how our naive negotiators could have expected that the Americans would *ever* agree to give Canada such exemptions.) Ultimately, as before the NAFTA agreement was signed, U.S. trade laws prevail – laws which they will certainly change at any time, and in any way, they wish. As in all other bilateral Canada-U.S. trade, U.S. domestic concerns will always win out.

As a result of the American refusal to obey the softwood lumber tribunal decisions, NAFTA has become a joke, and is effectively a dead agreement. As journalist Thomas Walkom has written, "Free trade is much less than it seems; the NAFTA emperor has no clothes."[5] Bear in mind that panel after panel decided that duties on Canadian softwood should never have been imposed by the Americans. The United States should have admitted as much and should have returned all of the $5-billion in duties unfairly and incorrectly charged to and collected from us, plus interest.

The Harper government's incomprehensible softwood surrender set an appallingly bad and dangerous precedent. Canadian lumber producers, instead of having free trade and guaranteed access to the U.S. market, now have their exports limited and a 15 percent export tax if prices "are too low" or "exports are too high." In signing the new "compromise best-ever agreement" (according to Stephen Harper), Canada has acknowledged that NAFTA is not only worthless for lumber companies, but *for all businesses in Canada*. Which Canadian industries will, in the future, opt to spend tens of millions of dollars in legal fees to appear before NAFTA tribunals, knowing that even if they win it will be meaningless?

When Canadians were asked, in September 2005 by the public opinion polling firm The Strategic Counsel, if Canada has been tough enough with the United States over trade, especially softwood lumber, 76 percent said we haven't been tough enough.[6] Not tough enough? We acted like a bunch of helpless wimps. Simply put, we should have told the Americans that we were going to cut off all of their Canadian supply of oil, natural gas, uranium, and electricity until they agreed to abide by the NAFTA tribunal decisions, and until they returned all of the $5-billion plus interest that they owed us. Sure, the Canadian cowards who have been managing our trade policy would quake in their boots at the suggestion that we take a strong stance with the Americans as a matter of principle, but it's long past time we ignored these spineless individuals and their kind once and for all.

Remember, only one day after the Harper government completely surrendered, and allowed the Americans to keep $1-billion that belonged to this country, a U.S. court ordered the Bush administration to pay back *all* of the $5.3-billion collected from Canadian lumber companies. The U.S. Court of International Trade ordered the full refund because the duties were clearly illegal. But Harper and friends had already settled for only $4.3-billion, saying he was "proud of what we have achieved" and that his government stood by the deal it negotiated.

There have been millions of words written about the softwood lumber debacle over many years. The best explanation comes from Gordon Gibson, Vancouver newspaper columnist and former leader of the B.C. Liberal Party and a man with a family background in the lumber trade. He wrote that, after our huge victory in the U.S. Court of International Trade, by summer, the duties would have been gone with the money-return order soon to be achieved.

Alas, there is also the political track. Just after the Tories won the election, they had a chance to recruit Liberal David Emerson. How to justify this? He was the softwood expert; we need him.

U.S. President George Bush soon picked up the phone and asked Prime Minister Stephen Harper if he wouldn't like to settle softwood, fast. He called us. After five years as president, he suddenly wants to settle?

Mr. Bush had good reasons, of course. Our legal fight was going against him. We finally had the U.S. in one of their own courts – and they were losing. In addition, a Montana senator's seat was hanging on softwood. So, let's see if we can't hornswoggle the Canucks.

No problem. The inexperienced Harper administration seized the chance to brag that in only a couple of months it had been able to fix an issue the Libs couldn't solve for five years. And it would validate Emerson's sleazy jump to the Tories. As a result, they bought a deal so loaded in favour of the Americans it was arguably worse than the one Martin had turned down earlier.

Gibson went on to explain the horrifying details of the deal:

Export taxes were to be imposed even higher than the old tariffs, and this has now been done. We were to be capped at 30 per cent of the U.S. market when the Liberals had negotiated 34 per cent. Sawmills are now closing in Eastern Canada, jobs lost in the thousands. There will be lots more.

The U.S. protectionist lobby is to be handed $500-million of our money to pay their lawyers and refill their coffers to attack us again. We will pay for our own thrashing, in a fight we would have won had our government had the guts to stand up to the Americans.

Bad deal? Never mind. On April 27, Mr. Harper told an astonished House of Commons the issue had been settled. At that very hour, American lawyers were filing papers to restart the legal process. The U.S. lied, and we said nothing. Without that betrayal, the very next day the final NAFTA decision would have kicked in and countervail duties would have ended at once.

Continuing the political track, industry holdouts remained – so many that in desperation last week, the two governments jointly appealed to the U.S. court to dissolve everything on the basis it had never happened. We stipulated the U.S. had never done anything illegal, destroying five years' worth of legal victories and our shield against future harassment. And yet, immediately thereafter came the "return the money" order from the CIT.

So now we have those ongoing duties and a gutted NAFTA, plus supervision of much our forest law by Washington. Kind of makes you proud to be an allegedly sovereign Canadian, doesn't it?[7]

We have read with a combination of fascination, amusement, and scorn the remarkable words of former Canadian ambassadors Allan Gotlieb and Derek Burney, trade negotiator Gordon Ritchie and Senator Pat Carney, suggesting that Canada retaliate against the United States for failing to live up to the terms of the FTA. Carney called the Americans "jackboot negotiators." Burney said what the United States was doing was "beyond the pale," it was using "the tactic of the schoolyard bully." For Ritchie, who was the deputy chief FTA negotiator, "when you're dealing with a bully, and the bully punches you, you should punch him back." The American refusal to accept the NAFTA panel's decision is "an egregious, shocking, dishonourable breach of their obligations," says

Ritchie. Even the continentalist Simon Reisman, original chief FTA negotiator, says, "We should certainly load the gun on retaliation.[8]

Hilarious, of course. No one in Ottawa, no one, has the courage to retaliate. This is what Gordon Ritchie had to say in the *Globe and Mail* about the Americans:

> A senior Commerce Department official has formally declared that when the administration finally loses its cases before the free trade panels, after exhausting all reasonable (and some highly questionable) legal tricks, it will simply refuse to pay back the money illegally collected. . . . This is indisputably in direct contravention of the NAFTA and amounts to nothing less than a unilateral abrogation of the central provisions of the free-trade agreement.
>
> Today, the lumber industry is the target. Tomorrow, we can expect these tactics to be applied to everything from energy to agriculture.[9]

Except, of course, the Americans would never do anything to interrupt the flow of oil, natural gas, electricity, and uranium from Canada. But, still, despite Ritchie's words, Stephen Harper et al caved in and agreed to a cowardly surrender.

And what about the Liberals? While in office, International Trade Minister Jim Peterson insisted that the United States adhere to the NAFTA panel rulings, but the ultimate response from the Paul Martin government was somewhere between limp and feeble. And let's remember that before Martin became prime minister, Jean Chrétien said NAFTA was such a bad deal he would insist that it be renegotiated, and if the Americans refused he would abrogate the agreement. He even voted against it in parliament. Then, once he became prime minister, he implemented NAFTA without amendment.

For Gordon Ritchie, the new developments were vitally important: "The American refusal to comply (with the softwood lumber decisions) tears the heart out of NAFTA."[10] He went on to say that Canada would not have signed the NAFTA agreement if it had thought that the United States would ignore its commitments in the agreement. Moreover Canadian exporters, far from being guaranteed protection against unfair application of U.S. trade laws, are now actually in a worse legal position than exporters from non-NAFTA countries.[11]

Despite all of this, Thomas d'Aquino and the other brilliant continentalist leaders of the campaign for both the FTA and NAFTA have called for NAFTA to be reopened and the dispute mechanism redesigned to fit "the new realities of North American integration." It really makes one wonder about the intelligence on display from our corporate leaders. From Day One of the FTA negotiations, many of us said repeatedly that the Americans would never surrender the right to impose countervailing, anti-dumping, or other duties when pressures from their own industries and those who write the big cheques to their congressmen are a factor. Norman Spector wrote, "On softwood lumber, Canadians are discovering that the 'binding' dispute resolution mechanism in NAFTA is anything but binding. That's what many opponents of free trade argued in the 1988 election. The U.S. still insists that its domestic law trumps binational panel rulings."[12]

Amazingly, for our Canadian continentalists and compradors, "deepening NAFTA" is now an urgent priority. And why would that be? The Conference Board of Canada explains that "increasingly, Canadian businesses seeking to expand into the U.S. market would rather build new facilities there than expand existing operations in Canada."[13] Clearly, those obtuse continentalists don't understand that it was the two trade agreements that created the very environment that would encourage business in Canada to prefer to locate in the United States.

So, all in all, what were the FTA and NAFTA really all about? Tariffs were already either non-existent or minuscule, and exchange rates were far more important for trade, and everyone knew that. Cross-border flows of goods and services were steadily growing and essentially unimpeded. So, then, what were the two agreements meant to achieve? Obviously, one major corporate goal was to get governments to lower taxes: "Either you lower taxes or we move to the Southern U.S."; or, in the United

States, "Either you lower American taxes or we move to Mexico." And you better get rid of all those silly rules and regulations while you're at it, or off we go to China and India. And certainly now's the time for major privatizations and for getting rid of those antiquated laws regarding foreign ownership and control. You better understand that under free trade, the market rules. Corporations will have much more power, and in dozens of important ways the hands of the government will be tied. Great!

Murray Dobbin, reviewing *The Ursula Franklin Reader,* writes:

> For almost 20 years, since the advent of the Canada-U.S. free trade agreement in 1989, federal governments have been deliberately diminishing the nation-state in line with the notion that these entities are somehow passé and must get out of the way of the new imperative: transnational corporations and the international corporations and the international institutions such as the World Trade Organization, the International Monetary Fund and the World Bank that facilitate them.
>
> Ursula Franklin's way of expressing this phenomenon is priceless. . . . She writes, "In my picture of what is going on, we are being occupied by the marketeers just as the French and Norwegians were occupied by the Germans. We have, as they did, puppet governments who run the country for the benefit of the occupiers. We have, as they did, collaborators. . . . We are, as they were, threatened by deliberate wilfulness by people who have only contempt for those they occupy and who see their mission to turn over our territory to their masters."[14]

What is to be done? Former Foreign Minister Lloyd Axworthy has suggested that Canada should seriously consider getting out of NAFTA and relying more on the World Trade Organization. (I am not a fan of the WTO, but we would be much better off going this way.) But *how* can we get out of the FTA and NAFTA? Well, both agreements have clauses allowing Canada, with six months' notice, to withdraw from the agreements without penalties. Axworthy, responding in 2005 to the U.S. behaviour in defiance of all the softwood lumber rulings, wrote:

> The reality is that we are dealing with an American political system currently steeped in the ideology of "empire." It recognizes few rules, adheres only to those treaties that are expedient to basic interests, and believes that the only political current that counts is the exercise of raw power.
>
> Now, any sector of the U.S. economy that feels threatened by competition can use the domestic system to impose penalties and engage in constant harassment – read softwood lumber, beef, steel.
>
> Let's face it: This is a painful and uncertain time in our relations with the United States.[15]

For Axworthy, we should not be "listening to the chorus of continentalist claptrap promoting more Canada-U.S. integration." For Michael Byers, widely respected professor of international politics at the University of British Columbia,

> It's time to stand up for Canada – by insisting that the energy provisions be removed from NAFTA, that international law rather than U.S. domestic law be applied to all trade disputes, and that the protections for foreign investors be substantially revised. In the meantime, whenever our southern neighbour reneges or procrastinates on its promises, Canada should refuse to capitulate.[16]

Jeff Faux, distinguished fellow at the Economic Policy Institute in Washington, D.C., puts it succinctly: "NAFTA protects the interests of large corporate investors while undercutting workers' rights, environmental protections and democratic accountability."[17]

Oh, yes, for all the NAFTA enthusiasts, the U.S. Coalition for Fair Lumber Imports has already announced that the United States will *never* accept a pro-Canada NAFTA ruling: "If necessary, our government will change the laws."

Stephen Harper proudly announced that the agreement he reached with the Americans regarding softwood put to rest the issue for many years into the future. But on June 16, 2007, a *Globe and Mail* article informed us that

> less than one year after Ottawa signed a deal that was sup-
> posed to end the Canada-U.S. softwood lumber dispute,
> fresh U.S. complaints are forcing it to consider concessions
> that could mean higher export taxes for some Canadian firms
> . . . concessions to Washington that could cost tens of mil-
> lions of dollars.

And by August 2007, the U.S. government, prodded by the Coalition for Fair Lumber Imports, launched formal complaints because Canada "violated" the terms of the agreement that brought Mr. Harper such pride.

Meanwhile, as Canada runs short of natural gas – which heats almost 50 percent of homes in this country and is the main source of energy for more than half of our manufacturing – because of NAFTA we must continue to supply the Americans with well over half of our entire production, like it or not.

In January 2008, U.S. officials said that under NAFTA provisions they expect to formally challenge Stephen Harper's $1-billion aid package for single-industry towns that have been hurt by plant closures resulting from the slowing U.S. economy.

I hope you will keep all this in mind when you read the conclusion of this book.

TRADE IN GOODS AND SERVICES

"A rear-view mirror approach to the world"

C anada prides itself as being a major trading nation. And indeed we are, but not quite as "major" as many Canadians assume. In fact, some of us have been saying for many years that while we certainly fully recognize the importance and the value of trade, far too many of our business leaders, and too many of our politicians, put too much emphasis on trade and not nearly enough on domestic production and on the upgrading of our resources.

Among the OECD countries, 14 have a higher level of trade as a percentage of their GDP than Canada. Measured this way, Canada is well behind such countries as Luxembourg, Belgium, the Czech Republic, and Ireland, is well behind the OECD average, and is even further behind the EU15 average.

What will surely surprise many trade advocates in Canada is that the supposed great trading nations, the United States and Japan, both have small trade-to-GDP percentages, Japan at only 12.2 percent and the United States at only 12.7 percent. No other OECD country comes anywhere near these very low figures. We'll look at Canada's percentage shortly.

Measured another way, when you compare Canada's total exports with the exports of other countries, in a list of the top 40 exporting countries, we're in ninth place. When you consider our exports as a percentage of total world exports, only the United States, Germany, the United Kingdom, Japan, France, China, Italy, and the Netherlands have a larger share. Canadian exports are well ahead of such countries as South Korea, Mexico, Russia, Sweden, Australia, Brazil, and India, for example.[1]

In 2006, Canada's share of world exports was 3.2 percent, slightly down from the previous 10-year average of 3.74 percent of all exports, and also down from 1994's 3.6 percent, the year we joined NAFTA.

While some rightly complain that Canada's share of world exports has been declining, the same is true for every G7 country. France fell from 6.3 percent in 1992 to 4.2 percent in 2006, Germany from 10.6 percent to 8.8 percent, Italy from 4.9 percent to 3.5 percent, Japan from 8 percent to 4.8 percent, the United Kingdom from 5.4 percent to 4.6 percent, and the United States from a high of 14 percent in 1999 to 10 percent in 2006. Total OECD exports fell from 76.7 percent of all exports in 1992 to 65.2 percent in 2006.[2]

Looking at per-capita exports, Canada was in 14th place in 2004, behind such countries as Switzerland, Norway, Belgium, Luxembourg, and the Netherlands, but ahead of the great trading nations such as the United States, Japan, the United Kingdom, and Germany.

Canada's merchandise exports to the world in 2006 were just over $455.7-billion – a record – but imports rose almost four times faster than exports, to a record $404.4-billion. This brought our merchandise trade balance down by over $11-billion to $51.3-billion, its lowest level since 1999, while the trade surplus with the United States came in at $96.1-billion, the lowest amount since 2003. If you measure imports and exports in chained[3] 2002 dollars, while exports exceeded imports by $50.9-billion in 2002, by 2006 imports exceeded exports by $39.9-billion for a huge negative change of $90.8-billion.[4] Canada's exports to the United States were 78.9 percent of all our exports, down slightly from 81 percent in 2005. Canada's total trade deficit with all other countries amounted to $43.94-billion. By 2007 exports to the U.S. had fallen to 75 percent of all exports.

In 2006, our leading exports were machinery and equipment, followed by industrial goods and materials, energy products, automotive products, forestry products, and agricultural and fishing products.[5]

Since 1971, Canada has always had a trade surplus in forestry, energy, and agricultural products. But as we've already seen, autos, which in 1999 had the largest trade surplus of any sector (more than $14.3-billion), went into a deficit in the summer of 2006. Auto exports reached a high of $97.9-billion in 2000, but declined to $82.5-billion in 2006, the lowest amount since 1998. Energy exports passed forestry exports in 2000, and the energy surplus reached a record $53-billion in 2005, mainly due to our increased natural gas exports. Forestry exports amounted to almost $43-billion in 2000, but fell to $33.3-billion in 2006.

Statistics Canada says that

> on average, resources have accounted for about half of our exports over the last 15 years. In 2005, the proportion jumped to 57%, with energy exports to the U.S. leading the way.
>
> In contrast to resources, exports of finished products fell sharply after 2000.
>
> There was a marked drop in the U.S. share of imports into Canada in recent years, unprecedented in the history of Canada-U.S. trade. The U.S. share of our total imports declined every year from their peak in 1997. In 2005, they made up only 56.6% of our total imports, the lowest since the 1930s.
>
> Since 2000, Japan has fallen behind China to third place on the list of our largest trading partners.[6]

In 2006, only machinery and equipment exports, at $94.7-billion, were greater than our rapidly growing energy exports, which increased from only $29.9-billion in 2000 to over $86.8-billion in 2006.

Most of our imports still come from the United States. Our biggest imports are, in descending order, machinery and equipment, industrial goods, automobile products, consumer goods, energy products, agricultural and fish products, followed by forest products at only just over $3.1-billion.

In terms of our manufacturing exports as a percentage of total exports, Canada is way down in 25th place among OECD nations, and well below the G7, EU15, and OECD averages.[7]

In recent years, the pattern of Canadian exports has changed dramatically. Heather Scoffield of the *Globe and Mail* writes, "Canada's trade surplus has become dependent on energy and commodity exports, marking a major shift in the structure of the country's economy and raising a worrisome question about its diversification."[8] Statistics Canada reports that "only record high surpluses in energy and industrial goods have sustained the trade surplus at a high level."

For Peter Hall of Export Development Canada, the decline into deficit in our auto sector trade is "ominous." Hall says the "high" dollar is taking a major toll on manufacturing, forestry, and auto exports.[9]

In March 2005, David Crane reported on a speech given by World Bank president Jim Wolfensohn, who expressed surprise that Canada, which likes to portray itself as a trading nation, sent only 6 percent of its exports to the huge markets of the developing world, which include China, India, Mexico, Brazil, and South Africa. Crane writes:

> Canada's business community – with a few notable exceptions – is pursuing a North American strategy, not a global strategy. It's an approach shared by think tanks like the C.D. Howe Institute and by strong enthusiasts for deep integration into the United States like Allan Gotlieb, Michael Hart and Thomas Courchene, as well as Tom d'Aquino and the Canadian Council of Chief Executives.
>
> But, this rear-view mirror approach to the world could cost Canadians a lot in the years ahead. It means that Canada could miss out on the vast economic opportunities that will emerge as populations grow and economies advance in the rest of the world.
>
> Canada's future well-being will depend on companies with a global strategy, not a North American strategy. But where are their voices to counter the intense propaganda

campaign by the deep integrationists to tie our future to the United States?[10]

From 1994 to 2004, our two-way trade with the European Union was less than one-10th of our total trade, but since imports from the EU increased substantially, our trade deficit increased to $11.2-billion in 2006. The annual rate of growth in our imports from the EU was second only to the rapid growth of imports from China. As our imports from China have risen dramatically, quadrupling over the past four years at a rate 60 percent faster than Chinese exports to the United States, Canada's perennial trade deficit with the rest of the world other than the United States is set to increase rapidly. In 2006 and 2007 our two-year trade deficit with China increased to some $54-billion. At the same time, China became the number one seller of goods to the United States, replacing Canada in that position.

Canada does poorly when it comes to trade in services. Services, as defined by the *IMF Balance of Payments Manual,* include freight and passenger transport, travel, postal, telephone, and satellite communications charges, construction, insurance, financial services, computer and information services, royalties, licence fees, cultural and recreational services, and professional fees. Twenty-four OECD countries have better service trade balances, and only four have worse. The United States, the United Kingdom, and, surprisingly, Spain have the largest service trade surpluses, the United States with a $47.8-billion (U.S.) surplus, followed by the United Kingdom with $39-billion (U.S.), and Spain with $27.6-billion (U.S.). Canada has had a service trade deficit every single year since 1950. In 2006, it was a record $15.2-billion.

The OECD average for the export of services is about 22 percent of the country's total exports. In Canada, it's only 12.5 percent.

While we very frequently hear about Canada's exports, we seldom consider imports into Canada. Import penetration, as defined by the OECD, "shows the extent to which the demand for goods and services is being met by foreign producers rather than from domestic production." In import penetration of goods and services as a percentage of total expenditure, in 2006 Canada stood at 29.7 percent. This is well above the OECD average of 21.1 percent. In comparison, the United States was at only 14.4 percent and Japan 9 percent.[11]

Of course, it's a given that Canada, with so much foreign ownership and control, and so many branch plant operations, will have a high level of non-arms-length imports from parent firms, mostly in the United States.

Philip Cross and Ziad Ghanem of Statistics Canada expand further:

> The importance of resources to our overall exports is often discussed, with a figure of 40% commonly cited. This share has risen to 50% of gross exports thanks to the commodity boom of the last two years. But subtracting out the higher import content of manufactured exports raises the share of resources to over 60%. This puts Canada in a unique class of major industrial nations, alongside nations such as Norway and Australia, where resource exports dominate. They are polar opposites of Japan, which imports most of its resources and exports almost none.
>
> Much of the growth of gross exports in the last decade reflected the increasing use of imported components, not higher value added exports. Value added exports, which include only inputs purchased in Canada are the key determinant of domestic output and jobs.

The two Statistics Canada economists show that the import content in our auto exports was 51 percent in 2004, in machinery and equipment exports it was 36 percent, and it was over 28 percent in industrial goods and consumer goods.[12] No exact current figures are available, but several reliable estimates suggest that 40 percent of Canada's trade with the United States is intra-firm.

Turning again to our trade with the United States, in 2006 our merchandise trade surplus was $96.1-billion. But our deficit with the rest of the world set a record of $43.94-billion, up from $34.65 in 2004. Overall,

Canada had a merchandise trade surplus well below the record surplus of $70.66-billion in 2001. While Canada has a large perennial trade surplus with the United States, we have had a deficit with the rest of the world almost every year, with the exception of a small surplus in 1995.

Here's something that will surprise a great many people. About two million Canadian jobs depend on our exports to the United States, while an estimated 5.2 million American jobs are dependent on their sales to Canada. Why the imbalance when we have such a large trade surplus? The answer is pretty simple. A very high proportion of our exports are non-labour-intensive resources or are finished products containing a high proportion of imported parts and components, while a very high proportion of U.S. exports to Canada are labour-intensive finished products originating in the United States.

In recent years, Canada has been the number one market for exports from 39 of the 50 U.S. states, and we were in the top three in another eight states. The United States sells more to Canada every year than it does to all the 25 EU countries put together, and Canada as a whole has been the number one export market for the United States since the end of the Second World War. U.S. exports to Canada represent 22.5 percent of all American exports.

Now for some completely irresponsible BS.

According to Frank McKenna, our former ambassador to the United States, our trade "relationship with the U.S. represents 40 percent of our gross domestic product. . . . We need to be respectful of it."[13] Prominent Canadian businessman and former civil servant Paul Tellier has said that "45 percent of Canada's economy is dependent on trade." Allan Gotlieb, another former Canadian ambassador in Washington, says, "Almost 40 percent of Canadian income depends on our access to the U.S. market." *Maclean's*, in their November 27, 2006, issue, repeated a Conference Board of Canada claim that our trade with the United States accounts for 72 percent of GDP and warned that "all aspects of our domestic economy are linked to our trading relationships with other countries. . . . If our borders were shuttered to international trade . . . three-quarters of our economy would grind to a halt."

Such claims range somewhere between pure garbage and total nonsense. The reality, supported by Statistics Canada research by some of their top people, places our net trade with the United States at some 20 to 23 percent of GDP. Moreover, to suggest that our borders might be shuttered is totally ridiculous considering the many millions of Americans whose income depends on their company's exports to Canada and the ongoing, increasing U.S. demand for Canadian resources.

If you wish to read more about the ridiculous claims of McKenna, Tellier, Gotlieb, and the Conference Board (among many others), I suggest you look at the excellent work on the subject by Statistics Canada's chief of current analysis, Philip Cross,[14] or at a copy of *Lies, Damned Lies and Trade Statistics: North American Integration and the Exaggeration of Canadian Exports* by economist Erin Weir.[15] In his first-class publication, Weir shows that

> The notion that international exports account for nearly half of Canada's economy significantly influences public-policy debates. However, the notion overstates the economic importance of trade flows by comparing gross exports, including a substantial amount of imported content, to value-added Gross Domestic Product (GDP), consisting only of Canadian content. The value of a product is counted in export statistics as many times as it crosses the border, but the value-added to it in Canada is counted only once toward GDP. Statistics Canada figures show that, in value-added terms, exports account for about a quarter of Canada's economy.

And of course exports to the United States would then only be some 19 to 20 percent of GDP in sharp contrast to the ridiculous claims of the likes of BMO Capital Markets chief economist Sherry Cooper, Allan Gotlieb, Brian Mulroney, and Thomas d'Aquino, among others already mentioned. All of these people should know better than their inflated claims, and probably do know better. But as I have written elsewhere,

"An absence of facts to back their ideology has never bothered our Canadian Americanizers."

Mind you, measuring trade as a percentage of GDP as a leading indication of the health of an economy is a mug's game. Witness, for example, that Singapore has a trade-to-GDP percentage of over 450 percent. Hungary, the Netherlands, and Taiwan are well over 100 percent. The many prominent Canadian journalists, politicians, and business leaders who so often claim that 45 percent of the Canadian economy is dependent on our trade with the United States are badly misinformed and invariably go on to misinform the Canadian public.

Looking at our trade picture overall reveals some worrisome developments.

- Imports are rising much faster than exports, particularly imports from China, the EU, and Mexico.
- Trade balances are declining. The trade surplus for 2006 of just over over $5-billion was the smallest in seven years.
- Trade deficits with the rest of the world other than the United States are increasing.
- There is a new and serious declining trade balance in the auto trade.
- There is a decrease in our exports of finished products, and a growing dependency on exports of energy and resources.

In 2004, Canada's export volumes of goods and services increased by 5.2 percent, but import volumes increased by 8.2 percent. In 2005, exports increased by only 2.1 percent, while imports increased by 7.1 percent. Then, in 2006, exports were up by 1.1 percent, but imports were up by 5.9 percent.[16] It's not a good trend.

GLOBALIZATION

INCREASING DOUBT AND RESISTANCE

I f globalization simply means greater commercial intercourse and better communication between the peoples of the world, who can be opposed to it? But if it also means something quite undemocratic which has a profoundly negative impact on the quality of life, then it's a different story.

A big change has taken place recently in how millions of people around the world look upon globalization. As every day goes by, the downsides of globalization are becoming more apparent, in the United States, in Europe, and in most developing countries. The prevailing right-wing global ideology is giving way to increasing doubt and resistance.

In the 1990s, officials from the U.S. Treasury Department, the International Monetary Fund, and the World Bank enthusiastically promoted lower trade barriers, deregulation, and the free flow of capital. But it's now clear that their prescription hasn't worked the way they claimed it would. In most countries, the gap between rich and poor has widened. More than two billion people still live on the equivalent of less than a dollar a day. The world's poorest countries are increasingly convinced that the rich countries have their own priorities first and foremost on their agendas, and simply don't care about the world in general.

Nobel prize winning economist Joseph E. Stiglitz has become an outspoken and articulate critic of our current concept of globalization. He points to the increasing inequality in a host of countries after trade was liberalized. Earlier in this book, you saw evidence of the increasing inequality in Canada, beginning after the FTA came into effect.

In a 2005 book,[1] Stiglitz and co-author Andrew Charlton argue that every developing country that went on to success initially protected its markets until it was ready to relax controls. They also argue that all World Trade Organization (WTO) members should commit to giving full access to the poorer developing countries. As former U.S. Secretary of Labor Robert B. Reich has written:

> Free trade is already disproportionately benefiting the best educated and best connected. The wealthy are growing much wealthier, while the middle class is being squeezed.
>
> Until gains are more widely shared – within richer countries as well as between richer and poorer – we can kiss any further round of trade liberalization good-bye.[2]

Perhaps the tide first began to turn for globalization with the rather unexpected repudiation of the proposed Multilateral Agreement on Investment, which would have allowed foreign ownership and control to increase virtually unchecked. Perhaps it was when nations such as Malaysia began to stand up so successfully to the rigid, conservative orthodoxy of the International Monetary Fund. Perhaps it was the steady accumulation of data showing how the rich were getting much richer while many of the rest were either standing still or getting poorer. Perhaps it was the arrogant, doctrinaire faith of the proponents of globalization that produced the growing hostile reaction.

Certainly, the rejection of the referenda promoting even further European integration at the cost of yet more national sovereignty, plus the collapse of the Doha Round of the WTO talks, and also of George W. Bush's proposed Free Trade Area of the Americas, are all clear indications of the trend John Ralston Saul describes in his excellent book *The Collapse of Globalism*.[3] For Saul,

The very idea of globalization is now slipping away. . . . Leading figures who once said nation-states should be subject to economic forces now say they should be reinforced. . . . Prophets of globalization who said, "Privatize, privatize, privatize," now say they were wrong because the national rule of law is more important.

Increasingly strong nation-states like India and Brazil are challenging the received wisdom of global economics. Pharmaceutical transnationals find themselves ducking and weaving to avoid citizen movements.

The conviction that citizens have such powers lies at the heart of the idea of civilization as a shared project. And the more people are confident that there are real choices, the more they want to vote . . . the more they want to become involved in society.

In recent years, the tide has turned quite dramatically. More and more developing nations are thinking of globalization as Western imperialism that produces forced privatization, more foreign ownership, more foreign control, mandatory cuts in social spending, and so on.

According to the 2003 United Nations *Human Development Report,* at the end of the 1990s, 54 countries were poorer than they were at the start of the decade. And the countries that were much better off, such as India and China, had very protected economies.

Many think that globalization is in fact working exactly the way corporate leaders intended. Workers' share of GDP is now lower than it's been in 30 years, and real wages have been, for the most part, stagnant or even decreasing. The real weekly earnings of a median American worker are down by 4 percent over the five years to 2006. At the same time, corporate profits, as we have already seen, are at or close to all-time highs as a percentage of GDP. In the United States, corporate profits almost doubled between 2001 and 2006. Meanwhile, the top 1 percent of American earners make about 16 percent of all U.S. income, double the share they made in 1980.

American economist Jeff Faux says that, with globalization, capital has left labour in the dust, creating a "global class war" between a "global governing class" and all others. Michael Hirst, senior editor at *Newsweek*'s Washington bureau, writes:

> Faux is clearly correct that the balance of power between labor and capital has shifted dramatically. . . . Wall Street has the whip hand over corporate performance . . . the gap between executive and worker pay has widened to record levels [and] even incompetent executives enjoy golden parachutes while high performing employees are laid off without apology.[4]

Meanwhile, "the American labor movement is a pitiful shadow of its former self, the victim of a 'China price' set half a world away by a seemingly limitless supply of cheap labor."[5]

As Noam Chomsky has pointed out, "Globalization used in a neutral sense just means 'international integration.'" And who could object to more of that, if it brought greater freedom and higher standards of living, just as who could object to "free trade" if it had a similar result? But Chomsky points out that the way globalization has evolved has meant the promotion of the "rights of investors, lenders, corporations, banks, financial institutions and so on. The people who now pretty much own the world have distorted the term 'globalization' to their extremely doctrinaire position . . . so there's even more concentrated wealth and power."[6]

A 2006 study by Gene Grossman and Esteban Rossi-Hansberg of Princeton University, shows that positive effects of globalization in terms of productivity were not enough to offset the downward impact on the wages of workers. For *The Economist,*

> In America, the Euro area and Japan, total wages have fallen to their lowest share of national income in decades, whereas the share of profits has surged. In their eagerness to applaud the benefits of globalization to economies as a whole, economists were strangely reluctant to admit that in recent years

the average real pay of rich-country workers has stagnated or
fallen.[7]

Two months later, the same conservative publication had this to say:

A Marxist would say this is a classic case of big business
exploiting the workers. Funnily enough, some out-and-out
capitalist economists would agree. They argue that profits in
America, for example, are at a 40-year high as a percentage
of GDP precisely because capital is winning at the expense of
labour. Globalization has brought the Indian and Chinese
workforces into the world economy, which has kept the lid on
wage costs.

Some will argue that increased GDP proves the success of globalization. But GDP figures by themselves can be misleading. If increased
GDP doesn't translate into a better standard of living and quality of life
for the majority, then are the built-in inequities of globalization so
harmful that alternate strategies must be considered? Increasingly, more
and more people around the world are coming to this conclusion.

One obvious and very important impact of globalization is that corporate tax rates have been driven down from averages of 37.5 percent in
1996 in the richer OECD countries to only 31 percent in 2003 and are
headed even lower now. Even more obvious is the lower share of total
income tax that corporations now pay in most countries, an OECD
average of only 9.6 percent in 2004.[8]

Meanwhile, the world's richest countries continue to maintain trade
barriers to the exports of some of the world's poorest countries. While
they provide aid to the poor countries, their enormous agricultural subsidies go far beyond the levels of all aid, destroying the markets and liveli-
hoods for producers in poor countries around the world.

Paul Hellyer, in his book *One Big Party*,[9] had this to say about globalization:

Under its benevolent wrapping it is a plan to strip elected representatives at all levels of government of their power to legislate on behalf of ordinary people, and to transfer that power to unelected, unaccountable international bureaucrats implementing rules laid down by the globalizers. It is a plan to end popular democracy as we know it, and substitute a plutocracy of the wealthy elite.

Not too different, I think, from the intent of both the FTA and NAFTA, and the plans of the gang of deep-integrationists that met at the Banff Springs Hotel, about which you will read more in the conclusion of this book.

PART SEVEN

FOREIGN AID

BREAKING PROMISES AND BREAKING HEARTS

"A scandalous betrayal"
"A disaster as avoidable as it is predictable"

Before we go on to examine Canada's truly dismal foreign aid record, let's consider what needs to be done and what adequate levels of foreign aid can accomplish. We all know about the almost 3,000 people who were killed in the tragic destruction of the World Trade Center's twin towers. But the world-famous American development economist Jeffrey Sachs says that every morning our daily newspaper should report: More than 20,000 perished yesterday of extreme poverty.

Or let me suggest a different headline: Twenty-eight thousand children under the age of five will die today of easily preventable causes.

Or, how about: Tomorrow, every 30 seconds, a child whose life could have been saved will die from malaria. In Africa, 3,000 children will die tomorrow from malaria, while many more will be left with lifelong brain damage or paralysis.[1]

Or perhaps: More than 6.5 million children will die this year from hunger, an average of some 18,000 every single day.

Jeffrey Sachs knows, in the words of *The New Yorker* writer John Cassidy, that "every year millions of children perish in infancy. In the face of this ongoing catastrophe, this year the American government will extend to poor countries . . . a tiny fraction of the Pentagon budget and about an eighth of one percent of the gross domestic product."[2]

Don Gillmore, in reviewing Sach's book *The End of Poverty,* writes: "In

2002, the U.S. gave $3 toward each sub-Saharan African. After taking out the administrative costs, consultant fees, emergency aid, and the money used to service debt, the donation amounted to 6 cents per person."[3]

Unicef, in their superb publication *Child Poverty in Rich Countries,* 2005, said that the proportion of children living in poverty in the developed world had increased in 17 of the 24 OECD nations for which data are available. Without question, "No matter which of the commonly-used poverty measures is applied, the situation of children is seen to have deteriorated over the last decade."

The 2006 Unicef Canada report said that some 10.6 million children will die globally each year from pneumonia, diarrheal disease, which kills 3,900 under the age of five every day, or 1.6 million children annually, and measles, which kills more than 500,000 children a year.

A few months after the Unicef study was released, it was reported that Canada, which, more two years earlier, had promised to become a major supplier of inexpensive drugs to Africa, had in fact failed to supply a single pill under the proposed program. When the aid was promised it was described as a major step of global leadership. Stephen Lewis, the former United Nations special envoy for HIV/AIDS in Africa, called it "a stunning breakthrough."[4] In June 2006, Lewis termed the global aid shortfall "a scandalous betrayal. . . . For heavens sakes, live up to your commitments." Finally, in August 2007, after lengthy disputes, a small breakthrough allowed shipments of an AIDS drug to go to Rwanda. Critics say that the agreement will likely not be repeated. Meanwhile, some 8,000 people die of AIDS in Africa every day.

Unicef Canada tells us that of all the deaths today of children under five years, the underlying cause of 53 percent is undernutrition. Many thousands of children die every day because of the lack of aid money and the lack of political will and public commitment. Eliminating or controlling the leading causes of child deaths is possible, but somehow the plight of vulnerable children is not on our politicians' radar screen.

Unicef reports that, in 2003, "only about 39 percent of the children suffering from acute respiratory infection in sub-Saharan Africa were taken to see a health provider. That means that some 14 million children in sub-Saharan Africa and South Asia were not treated."[5] Unicef goes on to say that "every year, measles is directly responsible for the deaths of a half million children. Even when it does not result in death, measles is often responsible for blindness, malnutrition and deafness." In addition, "by 2010 there will be up to 25 million children orphaned by AIDS in the world, with the majority living in sub-Saharan Africa."

In sub-Saharan Africa only 5 percent of those infected with HIV/AIDS have access to life-saving drugs.

Now consider this. Some 70 to 80 percent of Canadians believe Canada is "very generous" when it comes to helping poor countries.[6] But are we?

In both 2004 and 2005, in a list of 22 OECD countries, Canada was down in 14th place in donor assistance as a percentage of gross national income (GNI), far behind other countries, some of which contributed more than the stated and commonly agreed-to goal of 0.7 percent of GNI. In 2006, we slipped further down the list, to 16th place.

In 2004, despite so much hand-wringing and rhetoric year after year in Ottawa, Canada stood at a poor 0.34 percent of GNI. Sweden and Norway were at 0.94 percent, the Netherlands and Luxembourg were at 0.82 percent, and Denmark was at 0.81 percent. Bringing up the rear was the United States at only 0.22 percent, Portugal at 0.21 percent, and Greece at 0.17 percent.

In 2004 per-capita terms, measured in U.S. dollars, Canada gave $81, compared to $524 for Luxembourg, $477 for Norway, $377 for Denmark, and $302 for Sweden.

As Neil Brooks and Thaddeus Hwong have pointed out, "Nordic countries give an average of 0.71 percent of their GNI, more than double that of the Anglo-American countries."[7]

In May 2005, the European Union announced that its members would double their aid to poor countries by 2015, agreeing that by 2010 they would all be contributing at least 0.51 percent of GDP (which is slightly smaller, of course, than GNI), and that this would increase to the long-established target of 0.7 percent by 2015.

According to Oxfam, "The EU's bold announcement leaves the U.S. with nowhere to hide. If they fail to step up to the mark and pay their share, they will be responsible for derailing an historic deal on aid that would lift millions out of poverty."[8]

In January 2005, Jeffrey Sachs pointed out that the target of 0.7 percent of GDP represented about $150-billion (U.S.) a year, "half of which is from the United States alone. The missing aid would save millions of impoverished people this year from imminent death."[9] The United States, in 2005, spent 30 times as much on its military as it did on aid. And in June 2005, the U.S. Congress slashed U.S. aid by $2.5-billion, in a move supported by both Republicans and Democrats, leaving U.S. development assistance down 20 percent in real terms for 2006.

Meanwhile, polls show that a majority of Americans believe that U.S. foreign aid should be reduced.

And us? At one time, Canada was near the top of the list in our assistance to poor countries, although we never once reached the 0.7 percent target set by Prime Minister Lester Pearson in 1969. In 1975, we were at 0.5 percent of GNI. The year before Paul Martin became Finance minister, Canada's assistance as a percentage of gross national income was 0.46 percent. Then Martin went to work. The man whose plaintive pleas for the world's poor permeated his political career cut our aid all the way down to only 0.22 percent of GDP.

Late in 2005, the Irish rock star and anti-poverty crusader Bono, referring to Paul Martin's failure to live up to his promises regarding foreign aid, said, "I'm personally not just disappointed – I'm crushed." Earlier in the year, Martin refused to commit Canada to meeting the 0.7 percent Pearsonian goal even though, as mentioned above, five countries had long passed the objective, and seven others had promised to reach it by 2015 (the United Kingdom, Spain, Ireland, Germany, France, Finland, and Belgium). In June 2007, Bono said, "I can't believe that Canada has become such a laggard."

As Jeffrey Sachs and John MacArthur have pointed out, Canada's aid share to GDP

is actually much less than it was in the 1980s when Canadian incomes were roughly 20 percent lower than they are today.

Canada's shortfall vis-à-vis 0.7 percent (approximately $5-billion this year) would be enough to fund an entire global initiative to control malaria in Africa, a disease that needlessly kills more than two million children a year.[10]

The 1993 Liberal election Red Book promised 0.7 percent aid. Instead aid fell dramatically. So much for Liberal promises.

In the summer of 2006, Oxfam described the quantity of Canada's foreign aid as "abysmal." Moreover, some of our aid is "tied aid," which, as the University of British Columbia's David R. Boyd says, "forces recipient nations to purchase Canadian goods and services, often at inflated or uncompetitive prices."[11] The 2005 United Nations *Human Development Report* puts it bluntly: "Tied aid remains one of the most egregious abuses of poverty."

The United Nations estimates the cost of tied aid at $2.6-billion a year for low-income countries. According to the UN, "Much of what is reported as aid ends up back in rich countries, some of it as subsidies that benefit large companies. Perhaps the most egregious undermining of efficient aid is the practice of tying financial transfers to the purchase of services and goods from the donating countries."[12]

The United States tops the tied-aid list, with almost 85 percent of its aid tied. Canada is at more than 40 percent. In contrast, all the following countries are below 10 percent in tied aid: Switzerland, Belgium, Finland, Ireland, and the United Kingdom. Norway and Sweden are under 2 percent.

According to the United Nations, foreign aid from the well-to-do contributing countries is much lower now as a percentage of gross national income than it was back in 1990. Moreover,

Since 1990, increased prosperity in rich countries has done little to enhance generosity: per capita income has increased by $6,070, while per capita aid has fallen by $1.

Yet, for every $1 that rich countries spend on aid, they allocate another $10 to military budgets.

Currently spending on HIV/AIDS, a disease that claims 3 million lives a year, represents three days worth of military spending.

The $7 billion needed annually over the next decade to provide 2.6 billion people with access to clean water is less than Europeans spend on perfume. . . . This for an investment that would save an estimated 4,000 lives each day.[13]

The United Nations 2005 *Human Development Report* puts it this way:

Every hour more than 1,200 children die away from the glare of media attention. The causes of death will vary, but the overwhelming majority can be traced to a single pathology: poverty. With today's technology, financial resources and accumulated knowledge, the world has the capacity to overcome extreme deprivation. Yet as an international community we allow poverty to destroy lives. . . .

In the midst of an increasingly prosperous global economy, 10.7 million children every year do not live to see their fifth birthday, and more than 1 billion people survive in abject poverty on less than $1 a day.

The HIV/AIDS pandemic has inflicted the single greatest reversal in human development. In 2003, the pandemic claimed 3 million lives and left another 5 million people infected. Millions of children have been orphaned.

Meanwhile children die from malaria for want of a simple anti-mosquito bed net.[14]

As Nelson Mandela put it in 2005: "Massive poverty and obscene inequality are such terrible scourges of our times – times in which the world boasts breathtaking advances in science, technology, industry and wealth accumulation – that they have to rank alongside slavery and apartheid as social evils."

No indicator captures the divergence in human development opportunity more powerfully than child mortality. Had the progress of the 1980s been sustained since 1990, there would be 1.2 million fewer child deaths this year.

Over the next 10 years the gap between . . . [the United Nations' established] target and the current trend adds more than 41 million children who will die before their fifth birthday from the most readily curable of all diseases – poverty.

The UN projection for 2015 offers a clear warning: "To put it bluntly, the world is heading for a heavily sign-posted human development disaster. That disaster is as avoidable as it is predictable."

What could be done about the appalling numbers of children dying in poor countries? The 2005 *Human Development Report* is clear:

Child mortality is an area in which small investments yield high returns. Recent cross-country research on neonatal mortality identifies a set of interventions that, with 90% coverage in 75 high-mortality countries, could reduce death rates by 59% saving 2.3 million lives. The $4 billion cost represents two days' worth of military spending in developed countries.

What could be done? The *Human Development Report* sums it up precisely:

Measured in 2000 purchasing power parity terms, the cost of ending extreme poverty – the amount needed to lift 1 billion people above the $1 a day poverty line – is $300 billion. . . . This sounds like a large amount. But it is equivalent to less than 2% of the income of the richest 10% of the world's population.

What could be done? In 2003, military spending by OECD development donor nations amounted to $642-billion. Their official development assistance totalled only $69-billion. The United Nations puts it this way:

No G-7 country has a ratio of military expenditures to aid of less than 4:1. That ratio rises to 13:1 for the United Kingdom and 25:1 for the United States.

This 10:1 ratio of military spending to aid spending makes no sense. On any assessment of threat to human life there is an extraordinary mismatch between military budgets and human need.

Just about 3% of the increase in military spending between 2000 and 2003 could prevent the deaths of 3 million infants a year.[15]

What could be done? Since 2000, per-capita G7 military spending has increased by just under $170. Foreign aid spending has increased by only $11.

As I pointed out in my last book, *Rushing to Armageddon,* only 8 percent of current global military expenditures could properly fund universal programs in health, education, literacy, and AIDS reduction and substantially lessen the death of children and human suffering.

What could be done? In 2003, the United Nations Food and Agriculture Organization said that there is adequate arable land capable of providing everyone with a healthy diet and that would still be the case if the world's population increased as much as five times the current level!

What could be done? A *Globe and Mail* editorial has it right:

> The best way to bring the world's poorest countries out of their endless cycle of misery and reduce their dependence on foreign handouts is to bring down the barriers that block free access to developed markets for farm products, stop subsidizing farmers in wealthy countries, and end, once and for all, the egregious practice of dumping rice, cotton and agricultural goods in the Third World.[16]

From the president of Brazil: "Hunger is actually the worst of all weapons of mass destruction, claiming millions of victims every year.

Fighting hunger and poverty and promoting development are the truly sustainable way to achieve world peace . . . there will be neither peace nor development without social justice."[17]

In 2006, OECD foreign aid assistance fell by over 5 percent. For the University of British Columbia's David R. Boyd,

> It appears that the world's wealthy nations were just blowing smoke when they made their grand promises while thunderous speeches were delivered and great expectations created as the Millennium Development Goals to reduce the effects of world poverty were set.
>
> Official development assistance from the wealthy nations fell over 5 percent in 2006 and was forecast to fall again in 2007.
>
> For Canada the numbers are even worse. Canadian aid fell 9.2 percent in 2006 to a miserly 0.30 percent of gross national income, well below the OECD average.
>
> To make matters worse, Canada's proportion of administrative expenses (nearly 9 percent) is the second highest in the OECD, behind only Greece, and more than double the average, and Canada's proportion of tied aid is among the worst in the entire OECD, at more than 40 percent.[18]

The Harper government's 2007 budget promised an increase in our foreign aid, but an increase that would leave us mired down at only 0.272 percent of GDP for the 2007/2008 fiscal year. In late 2007, with great fanfare Harper announced new foreign aid initiatives "to save a million lives." But, it was simply a repackaging of assistance provided to UNICEF every year, with no increase in Canada's overall foreign aid.

According to Boyd, "Canada is breaking promises and breaking hearts."[19]

Bob Geldoff, speaking during the June 2007 G8 meetings, which discussed development assistance, said, "Canada is blocking a meaningful communiqué from the meetings. I think it's a shame for Canada to take that role."[20] And at the same time Jeffrey Sachs said, "Canada is nowhere to be found on this commitment at all. . . . There's no life in Canada's efforts. It's a huge surprise for those of us who believe in Canada's role. . . . We don't hear Canada's voice on major issues."[21]

Makes one really proud to be a Canadian, doesn't it?

In October 2007, Unicef reported that aid was indeed having a positive impact, with the number of child deaths falling below 10 million, to 9.7 million, in 2006, while under-five mortality rates were down by over 50 percent from 1960. There was much rejoicing when Unicef released the new estimates, especially when so much of the improvement came from relatively low-cost solutions such as mosquito netting. This said, the death of 9.7 million children in a single year is a terrible tragedy that could have easily been prevented.

I'll leave the final words in this chapter to Stephen Lewis:

> The ongoing plight of Africa forces me to perpetual rage. . . .
> I have spent the last four years watching people die. . . . Whenever I travel in Africa today, it feels as though everyone is hungry – hungry to the point of starvation. Hut after hut yields a picture of a mother, usually a young woman, in the final throes of life.
>
> The AIDS pandemic has taken a devastating toll. . . . I was visiting the adult medical wards. There were two people in every bed, head to foot and foot to head, and in most instances under the bed on the floor, each in an agony of full-blown AIDS.
>
> On the ten-hour night shift, to take care of sixty to seventy patients – every one of whom would have been in intensive care in a Canadian hospital – there would be one nurse.[22]

You might want to keep all of this in mind when you read the next chapter.

DEFENCE, THE MILITARY, THE ARMS TRADE, PEACEKEEPING, AND THE ARCTIC

"Dying in Afghanistan for a cause we cannot win"

I hope you will be able to read this chapter right after you have read the preceding chapter on foreign aid.

In 1985/1986, defence spending in Canada passed $8-billion for the first time. By 2005/2006, it had increased to over $15-billion, and it will be $18.24-billion in 2007/2008, the 13th highest in the world.

For years, many critics, the military, the defence contractors, the Americans, and many Canadian academics and journalists have complained that Canada wasn't spending nearly enough on defence, or perhaps what can be more accurately categorized as the military.

Current projections are that defence spending will soon surpass $20-billion and exceed Cold War levels. Steven Staples, who heads the Rideau Institute on International Affairs in Ottawa, says that today,

> Most of the money is being spent on American – not Canadian – military priorities to allow our forces to become much more capable of performing war-fighting missions with the United States.
>
> This transformation from a peacekeeper to a war-fighter is happening almost without any public debate at all.
>
> The Americans have continued to strongly pressure Canada into increased military spending.

For example, he says, "Canada spent $1 billion on old British submarines only because the Americans wanted us to do so, so they could use the diesel-electric subs, which they no longer had, for target practice."[1] Hard to believe, but true.

It's interesting to note that a 2005 public opinion poll showed that almost 70 percent of Canadians said that they wanted the same or a more distant military relationship with the United States. Very few Canadians wanted a closer relationship.

In its 2005 budget, the Martin government promised to increase military spending by $12.8-billion over five years. That fall, the U.S. journal *Defense News* said that Canada was already the 14th highest military spender in the world and the seventh largest military spender among the 26 NATO members, only marginally behind Germany as a percentage of GDP, although for years we had been near the bottom of the list in the ratio of spending to GDP, and well below the NATO average. Mind you, many NATO countries have regional conflicts and historical reasons for their high military spending; for example, Greece, at 4.1 percent of GDP, and Turkey at 4.9 percent.[2]

Canada is now 10th on a per-capita basis, and our per-capita spending is over twice the world average, while we've recently moved up to the sixth highest military spender among the 26 NATO countries[3] and, as Steve Staples has shown, we now outspend the lowest 12 NATO members combined.

And please note, not included in our defence or military budget is well over $10-billion spent each year for "security."

Even with our worrisome and controversial presence in Afghanistan, public opinion polls continue to show that Canadians place increased military spending well down the list of their priorities, which are invariably topped by the environment, health care, post-secondary education, poverty, child care, and more recently the economy.

For yet another indication of how out of touch our Canadian Senate often is, in 2005 Senator Colin Kenny of the Senate's National Security and Defence Committee called for defence spending to be increased to

$25-billion, with a further target of a huge $35-billion a year by 2012. The senators on this committee are some of the same senators who constantly appeal for even more and even larger tax cuts.

In the millions of words written about Canada in Afghanistan, I think it's difficult to top Thomas Walkom's perceptive column in the *Toronto Star* about Colin Kenny and his Senate defence committee's 2007 report.

> There is a bizarre disjunction in the Senate defence committee's useful – and remarkably frank – analysis of Canada's military role in Afghanistan. It's as if the 11 senators on the committee, having successfully outlined the near insurmountable problems associated with the Afghan war, couldn't bring themselves to accept the logical conclusion of their own analysis.
>
> On the one hand, their 16-page report convincingly paints a picture of a war that cannot be won. The Afghan government of President Hamid Karzai, it states bluntly, routinely shakes down its own citizens. Its army and police are, in the words of committee chair Colin Kenny, "corrupt and corrupter."
>
> But . . . optimism is hard to square with reality. In fact, they say, Canada's military presence in the southern province of Kandahar has not made the lives of Afghan citizens any better. It has made them worse. "Life is clearly more perilous because we are there," the report concludes.
>
> As for what Foreign Affairs Minister Peter MacKay calls Afghanistan's advances in women's rights, democracy and education, the senators are dismissive. Afghanistan, they conclude, is a medieval society that "does not want to be rebuilt in Canada's image." To suggest that our efforts will somehow miraculously create a modern, liberal state is to engage in the grossest kind of illusion. . . .
>
> They quote one former Canadian ambassador as saying that it would take five generations to make a difference in Afghanistan. They cite Lt.-Gen. Andrew Leslie, commander

of Canadian land forces, as saying the Afghan military mission alone would take 20 years. . . .

In the end, the senators can't quite bring themselves to accept the inescapable conclusion of their own hard-headed assessment. Everything about their report screams that Canada has no chance of success in Afghanistan. But clearly, the committee is unwilling to take the final step and call for the troops to come home.[4]

By the summer of 2007, public opinion polls in Canada were showing that an increasing number of Canadians felt we should withdraw our troops from Afghanistan before their mandate ends in February 2009. Many said Canada has had to bear too much of the most dangerous burden of the NATO mission. Many others said that the Harper government had failed to adequately explain the mission to Canadians.[5] By July 2007, only 40 percent of Canadians supported our military presence in Afghanistan.[6]

As the pressure built and opposition in Quebec became overwhelming as troops from that province were scheduled for deployment, Stephen Harper completely reversed his long-established and unequivocal position and announced that Canada's military mission would end in 2009 unless there was strong support in Parliament. Harper made this promise knowing full well that no such support was possible. Nevertheless, the pro-American "bipartisan" committee Harper appointed, headed by John Manley, will almost certainly have come out in favour of a continuing Canadian presence in Afghanistan. As the *Toronto Star*'s James Travers has written, "It will be a seismic shock if the handpicked panel recommends anything else."[7]

Before we go on, we should note that the Harper government has plans to spend an enormous $20.2-billion for planes, helicopters, ships, and trucks for the military.

Very briefly, let's look at Canada's role in the arms trade. In contrast to our self-image, Canada is the sixth largest supplier of military goods to the world, the third largest supplier of arms to the United States, behind only the United Kingdom and Italy, and our military exports have more than tripled during the last seven years.

The UN says the United States supplies the world with 30 percent of conventional arms, but the Stockholm International Peace Research Institute says that it's somewhat higher, and puts Russia in second place at some 30 percent. France is at 9 percent, Germany 6 percent, and the United Kingdom 4 percent. All of the following countries are at approximately 2 percent each: Sweden, Canada, the Netherlands, Italy, Greece, India, Japan, and Israel.[8]

One of the strongest and most effective lobbies in Canada is the Canadian Association of Defence and Security Industries. Some of the companies involved in the lobby group include SNC Technologies, which sells large quantities of ammunition to the U.S. military, General Dynamics, and Raytheon Aircraft. The Canadian industry's sales are about $7.5-billion a year. According to *Maclean's,* "In recent years Canadian military products have gone to over 70 countries including Egypt, Saudi Arabia, China and Venezuela."[9]

As B.C. journalist Stephen Hume has pointed out,

> By some estimates, the world spends more on weapons annually than the total yearly income for near half the world's population.
>
> Never in human history has the world been a more dangerous place for individuals – particularly for women and children – and the younger the children, the greater the risk.[10]

Now to peacekeeping. Over 50 years ago, a Canadian idea led to the founding of the first United Nations peacekeeping force to deal with the Suez crisis. Generations of Canadians know of, respect, and are proud of our role as peacekeepers. But without most Canadians being aware, all this has changed.

In 1991, Canada had almost 1,150 soldiers directly involved in UN peacekeeping operations. By the fall of 2008, we were down to only 57 out of a total of over 73,000 UN peacekeepers. Compared to the total number of military personnel other countries contribute to UN peace-keeping, we're now 36th in the world. Thirty-sixth! Where once we had one of the best and most admired peacekeeping forces in the world, today our military has been transformed and our reputation sadly diminished at a time when the UN needs our peacekeeping support more than ever before. In the words of well-known Canadian Major-General (ret'd) Lewis MacKenzie, "Somebody put a marker up and said, 'Rest in peace, peacekeeping,' because it is no more."[11] *The Economist* says the notion that Canadians do peacekeeping has become largely a myth.[12]

In 2006, there was a record high number of soldiers and police in UN peacekeeping operations. Where traditionally Canada's contribution was around 10 percent of UN forces, we're now down to only about one-10th of 1 percent of all those in the 16 UN missions (2006). Yet an October 2005 poll showed that 69 percent of Canadians still considered peace-keeping a defining characteristic of Canada.[13] Unfortunately, it is indeed, but not exactly the compliment Canadians thought they were expressing.

More and more Canadians (witness the trend in the public opinion polls) have realized that what we've been doing in Afghanistan (in a NATO, not a UN, operation) has much more to do with offensive military operations than with peacekeeping.

The UN's 2005 decision that the "responsibility to protect" should be a principle for future operations was strongly promoted by Canada as a basis for the protection of civilians. As Lloyd Axworthy and many others have argued for two years, Canada should be "relocating resources now dedicated to war-fighting in Afghanistan to peace-building initiatives that cry out for attention and leadership. Darfur leads the list."

In a 2007 public opinion poll, two-thirds of Canadians mistakenly agreed that "Canada is an essential contributor to peacekeeping,"[14] and three out of five said Canadians "are dying in Afghanistan for a cause we cannot win."[15]

James Travers writes of the persistent doubts about what could be reasonably achieved in Afghanistan and "the notable absence of political candour" from both the Conservatives and the Liberals. Travers says, "The military, along with the arms lobby, will howl, but one alternative is peacekeeping, and there's lots of it to be done. It's also true peacekeeping is not what the military wants to do."[16] As Michael Valpy has written,

> Canada has turned down so many United Nations' requests to join peacekeeping missions during the past decade that the UN has stopped asking.
>
> Today, there is, in fact, not a single Canadian officer in the UN's peacekeeping headquarters.[17]

By 2007, over 200,000 men, women, and children had been killed in the Darfur area of Sudan, and some 2.5 million had been made homeless by the conflict there. In August 2007, the United Nations finally decided to send a force of 26,000 military and police peacekeepers into Darfur. At this writing, there was no sign that there would be any Canadians among them.

Before we go on, a few words about U.S. military spending, which accounts for almost half the entire world's trillion-dollar-plus military expenditures. Depending on whom you choose to believe, U.S. "defence" spending is between six and 15 times Russia's, and between six and 10 times China's.

By early 2007, China had become the world's second biggest military spender, reportedly spending $49.5-billion annually, ahead of the United Kingdom, Japan, and Russia, and its spending was scheduled to increase by 18 percent during the year. As Douglas Roche, Canada's former ambassador for disarmament, has pointed out, all potential U.S. enemies put together spend only about 25 percent as much on their military spending as the United States, and U.S. spending is about 26 times the combined total of seven "rogue states."

By 2006, U.S. military spending was greater than all U.S. medicare and social security spending combined, coming in at some $529-billion (U.S.), a figure which included the costs in both Iraq and Afghanistan. It also includes about $73-billion for R&D for developing a whole new generation of deadly weapons.

In February 2007, President George W. Bush proposed a new military budget that would bring the total to $622-billion. And in August, the United States announced a staggering increase in the supply of American arms to the Middle East totalling $63-billion over a decade, including $20-billion in weapons sales to Saudi Arabia, which from 1990 to 2000 bought $40-billion worth of military equipment from the United States.

So much for the needs of poor Palestinians.

A few words about the Arctic. Who owns and controls Canada's Arctic waters? Canadians do, of course! Not so, say the Americans.

The United States has never recognized Canada's claim to sovereignty in the Arctic, regardless of what your maps show. And with global warming and the inevitable opening of the Northwest Passage to shipping, the Americans contend they don't need Canadian permission to sail through the Arctic islands, and that means no one needs permission, not Koreans, not Iranians, not terrorists, not anyone. The Americans say the Northwest Passage is "an international channel for passage" and they plan to expand the presence of the U.S. Navy in the region. The U.S. ambassador to Canada, David Wilkins, says, "the Northwest Passage is a strait to be [open] for international navigation. It's not a Canada-U.S. issue, it's a Canada-versus-the-rest-of-the-world issue."[18]

As well, the European Union considers the passage "neutral territory," while in 2007 Russia announced plans to vastly expand its territorial claims in the Arctic, well into Northern Canada and areas long thought to be rich in minerals, oil, and natural gas. Some estimates suggest that as much as 25 percent of the world's undiscovered hydrocarbons are in the Arctic.[19]

What has been the Harper government's response? They have cancelled their election-promised plans to build three armed military icebreakers and put in place strings of underwater sensors. Instead, they plan six to eight smaller patrol ships, while both Russia and the United States operate heavy icebreakers capable of sailing in the Arctic ice in any month. To the Conservatives' credit, in August 2007 the Harper government also announced plans to build two new military bases in the Arctic, a training centre and a deepwater port at the northern end of Baffin Island, a strategic spot near the eastern mouth of the Northwest Passage.

Peter Wilson of the Nunavut Planning Commission comments that Harper's

> "slushbreakers" won't actually be capable of operating in the Arctic year-round. They'll have to retreat south when it gets too cold for them up North [while] the U.S. and Russia have the capability to cruise through or under Canadian Arctic waters in any season.
>
> This is an embarrassment. There are many important things that Canada can do in its Arctic – all of them assert sovereignty.[20]

For President George W. Bush, "the North Pole is our backyard. The U.S. has huge geopolitical interests in the Arctic."[21] Not to mention the interests of the huge American oil companies.

For the Pentagon, Canadian claims of Arctic sovereignty are "excessive" and "tenuous." For the current U.S. ambassador to Canada, his country has long been clear on the matter: the Northwest Passage is an international strait and American ships do not need Canada's permission to pass through it. Many major maritime nations agree. For them, the passage represents a 7,000 to 8,000 kilometre shortening of the key trade route between Europe and the Pacific Ocean. For the United States, aside from strategic military concerns, the passage will shorten the sailing distance between Tokyo and New York by 7,000 kilometres.

The most recent estimates suggest that almost all of the summer sea ice will be gone no later than 2040, but massive melting will continue to take place much earlier. One study says that an area of summer ice the size of Alaska has melted over the past 28 years.[22] As University of British Columbia international law expert Michael Byers has pointed out,

> The briefing book given to Gordon O'Connor when he became defence minister states "If the current rate of ice thinning continues, the Northwest Passage could be open to more regular navigation by 2015."

International shipping in the Arctic entails serious environmental risks. An oil spill would cause catastrophic damage.

Then there are the security concerns. Ships carrying illicit cargoes could be attracted by the relative absence of a police or military presence. Smugglers, illegal migrants, even terrorist groups could regard an ice-free Arctic as an open backdoor to Canada and the U.S.[23]

Meanwhile, a new U.S. Navy estimate suggests that the Northwest Passage will be open to shipping for a minimum of one month every summer by as early as 2011, and a May 2007 report said that the Arctic ice has been melting 30 years faster than previously forecast. In September 2007, experts said that they were "stunned" by the rapid loss of ice and the accelerated loss every year since 2002. Since September 2005, the extent of Arctic sea ice has fallen by an enormous 600,000 square kilometres.

Stephen Harper promised unequivocally to build three armed icebreakers and an underwater sensor system in the Arctic. As for Canadian presence in the Arctic, Harper said, "Either we use it or we lose it."

Yet, at this writing, Byers points out,

> Harper's indecisiveness is all the more unfortunate because U.S. interests have changed. During the Cold War, the U.S. was focused on maintaining open access for its navy, and especially its submarines. Today, Washington is more concerned about terrorists sneaking into North America, or rogue states using the oceans to transport WMD.
>
> It cannot benefit the U.S. to have foreign vessels shielded from reasonable regulations and scrutiny by maintaining that the Northwest Passage is an international strait.[24]

Even the often objectionable and arrogant former U.S. ambassador Paul Cellucci agrees, telling an October 2006 foreign affairs conference in Ottawa that it would be preferable for the United States if Washington allowed Canadian control of the passage, instead of demanding that it be considered an international waterway open to shipping from all countries.

Unfortunately, yet once again, after years of both Liberal[25] and Conservative promises about Arctic sovereignty, only relatively small steps have been taken. You may remember that in 1986, when the U.S. icebreaker *Polar Sea* sailed through the Northwest Passage without seeking permission from Canada, I "bombed" the huge ship from the air with Canadian flags and "Turn Around and Go Home" notices, much to the embarrassment of Brian Mulroney and External Affairs Minister Joe Clark. This was a year after Clark, in the House of Commons, said, "The Arctic is not only a part of Canada, it is part of Canadian greatness. The policy of the Canadian government is to preserve the Canadian greatness undiminished. Canada's sovereignty in the Arctic is indivisible. It embraces land, sea and ice."[26]

It appears that Stephen Harper may continue to disappoint us in the same way his predecessors did.

An important point that has been largely missing in discussions in Canada about Arctic sovereignty was identified in a *New York Times* editorial:

> The United States does not find itself in a strong position. Misplaced fear among right-wing senators about losing "sovereignty" has kept the Senate from ratifying the Law of the Sea even though the United Nations approved it 25 years ago. This, in turn, means that the United States, with 1,000 miles of coastline in the Arctic has no seat at the negotiating table.[27]

By November 2007, because of the potential opening of the Northwest Passage, there was pressure mounting in the U.S. Senate for the country to change its stance and finally join the Law of the Sea.

Two last points for this chapter. In an excellent column in the *Globe and Mail*, Lawrence Martin criticized Stéphane Dion and the Liberal Party for being "strangely mute on the subject of Iraq and squandering a major political advantage where the Liberals have their [Conservative] opponent at their mercy." What, Martin wondered, was holding Dion back?

How can he stay silent? He should be siding with the war's more vocal Democratic naysayers in the U.S., and with the American people themselves.

The biggest, most politically damaging link between Canadian and American leaders is the Iraq bloodbath. The Liberals have Mr. Harper cornered and won't take advantage.[28]

Had it been up to Stephen Harper, Canadian coffins would have been returning from Iraq for years now, and for who knows how long into the future. For months, he argued vehemently in the House of Commons that Canada had a duty to stand with the United States in its war against Iraq.

Dion's reluctance to criticize forcefully George W. Bush and the people around him on the Iraq fiasco is very difficult to understand. It is indeed a squandered opportunity, both politically and a chance to do the popular right thing and show leadership in a sad and tragic vacuum.

I can't leave this chapter without strongly suggesting that you look at the Middle Powers Initiative website (www.middlepowers.org) for the important work of Canada's former Ambassador for Disarmament Douglas Roche. In particular, see his excellent May 31, 2007, testimony to the House of Commons Standing Committee on Foreign Affairs and International Development regarding de-alerting nuclear weapons, a Fissile Materials Cut-off Treaty, an effective Comprehensive Nuclear Test Ban Treaty, and other matters relating to the terrible dangers of nuclear proliferation. Roche notes that of 31 votable UN nuclear disarmament resolutions in 2006, the United States cast the sole No vote 12 times.

Nobel laureate Jody Williams visited Canada in late November 2007 to celebrate the 10th anniversary of the treaty to ban land mines, the treaty that Canada was so instrumental in promoting. But this time, she asked "Where's Canada's leadership in global issues right now? . . . You are really taking a back seat and it's really hard to understand."

PART EIGHT

GOVERNMENT IN CANADA

"Canada is the highest spending country in the world."
— STEPHEN HARPER, ANNOUNCING HIS INTENTION
TO SEEK THE LEADERSHIP OF THE CONSERVATIVE PARTY,
JANUARY 12, 2004

The best thing you can say about this statement is that it was nonsense then and it is nonsense now. Either Harper simply didn't know what he was talking about, or he was intentionally misleading Canadians. But Harper is not alone. If we believe the steady stream of propaganda coming from our far-right-wing Canadian plutocracy, most of our so-called "think tanks," some of our business press, and all of our ultra-conservative politicians, government spending in Canada has been out of control and is far greater than in most other countries.

Nonsense and more nonsense.

If we look at all government spending, central, provincial, and local, and include all social security expenditures, of the 30 OECD countries 17 spent more as a percent of GDP than Canada in 2006. At 39.5 percent in 2005, government spending in Canada was far below the Euro area average of 47.1 percent, and below the OECD average of 40.4 percent.[1] In 2007, total government spending in Canada was expected to fall to 38.1 percent of GDP.

And, of course, some of the countries that spend less than Canada have poor to appallingly bad social programs, countries such as Korea and the United States, for example.

It's interesting to note that, contrary to conventional wisdom, U.S. government spending, at 36.6 percent of GDP in 2005, isn't much less than our government spending in Canada. But government spending in the United States that year included over $500-billion for military and defence expenditures, plus some $220-billion to service the enormous U.S. federal debt.

So, quite contrary to what Stephen Harper told us, Canada – down in 20th and 18th place in 2005 and 2006 when total government spending is measured as a percentage of GDP – is among the lowest government spenders in the OECD.

In 2004, when Harper made his misleading claim, Canada's total government spending amounted to 39.4 percent of GDP, below the 40.4 percent G7 average, and well below the spending of France (54.4 percent), Italy (48.6 percent), Germany (47.7 percent), and the United Kingdom (44.1 percent). While the United States (36 percent) and Japan (37.3 percent) were below Canada's spending, most Canadians would be very unhappy with the poor government social programs in these two countries.

Moreover, Canada's string of 11 consecutive years of overall government surpluses (from 1997 to 2007) compares most favourably with the 12 consecutive years (from 1993 to 2004) of deficits in Japan, and the American post-Clinton string of massive, record-breaking deficits and enormous accumulated debt. In contrast, all Canadian governments, municipal, provincial, and federal, had total combined net financial liabilities of only 26.6 percent of GDP in 2006, compared to the G7 average of 50.0 percent, and liabilities in Italy of 94.6 percent, in Japan 85.4 percent, Germany 51.9 percent, France 42.5 percent, the United Kingdom 39.7 percent, and the United States 36 percent.[2] In 2005, per-capita government total debt in Canada came to $28,390. In the United States it was $48,700 (Canadian).[3]

In 2006/2007, the consolidated surplus for all Canadian governments and the Canada and Quebec Pension Plans was the highest on record, some $47.7-billion. There hasn't been a combined government deficit in Canada since 1998/1999. In 2006, by contrast, Washington recorded a deficit of an enormous $811-billion.

Two thousand and six was the fourth consecutive year in which Canada was the only G7 country to record a surplus, and this was also expected to be true for 2007 as well.

In 2005/2006, the largest surplus in Canada, some $13.2-billion, belonged to the federal government. This was Ottawa's ninth consecutive surplus. Next was Alberta's $7.3-billion surplus, followed by the Canada Pension Plan surplus of $7-billion. In 2006/2007, Ottawa's surplus grew to $13.8-billion.

Six provincial and territorial governments had deficits, and local governments had combined deficits of $2.9-billion.

In 2006/2007, consolidated provincial, territorial, and local governments recorded a surplus of $7.4-billion. Alberta's surplus increased to $8.5-billion.

Because of Ottawa's surpluses, federal debt charges fell to 15 cents out of every dollar of government revenue, while provincial/territorial charges were down to nine cents of every revenue dollar. Federal debt as a percentage of GDP peaked at 68.4 percent in 1995/1996, was down to 32.3 percent in 2006/2007, and is expected to be down to 29.7 percent in 2008/2009. By 2006/2007, the debt was down over $95.6-billion from the $562.9-billion peak and is now at its lowest level to GDP since 1981/1982.

These federal government surpluses are in sharp contrast to the federal deficits of over $30-billion which were common from the 1980s until 1995, the highest being just under $38-billion in 1985, the first full year of the Mulroney government.

Despite all of this, conventional right-wing "wisdom" is that government spending in Ottawa has been out of control and has been rising for years. But in fact the federal government's budgetary revenues have fallen, from 18.2 percent of GDP in 1997/1998 to 16.3 percent in 2006/2007, and are projected to be down to 15.3 percent in 2008/2009. Total federal Canadian government spending, as a percentage of GDP, has declined significantly in recent years. In 1982/1983, it amounted to 25.4 percent of GDP, but it is expected to be as low as 15 percent of GDP in 2008/2009.

Let's turn now to look at all of Ottawa's program spending as a percentage of GDP.

From the time the provinces and territories joined medicare in 1972 to 1979, Ottawa's program spending averaged 17.8 percent of GDP, and during the 1980s it was about the same, at 17.6 percent of GDP. But with the Americanizing impact of both the FTA and NAFTA, program spending in the 1990s fell all the way to 14.5 percent of GDP. By 2005/2006, it was down to only 12.7 percent, and at this writing the Conservative government forecasts it will be 12.9 percent for 2008/2009, well below pre-medicare days. The last numbers are far below the averages of the past few decades, but if you believe the likes of the Fraser Institute, et al., the Conservative government has gone wild and abandoned all the right-wing fiscal principles.

Dr. Janine Brodie of the University of Alberta has calculated that during the past decade Canadian governments have collectively reduced program spending from 41 percent to 31 percent of GDP.[4] In 2008/2009, Ottawa's transfers to persons (see below) will have sunk to 3.8 percent of GDP from a high of 5.9 percent in 1992/1993.

By 2006, a 5 percent drop from the pre-medicare program spending levels represented a drop in federal government spending of a colossal $72-billion.

I'm not suggesting for a moment that we should be spending $72-billion more on various government programs, but as we've already seen in the chapter on social spending, the comparatively low levels of program spending are so far below historical averages that they are incomprehensible when current demands regarding health care, post-secondary education, the environment, child poverty, our military, public housing, and so on, need to be addressed.

Some people think our very low level of program spending is great. Suggest to them that they read the chapters on poverty, social spending, welfare, health care, and education in this book.

To make matters worse, the Harper government's projections of program spending show a further decline to only 12.6 percent of GDP by 2011/2012.

Let's turn now to Ottawa's total spending, including transfers to persons and transfers to other levels of government, all program expenses including defence, and all interest payments on the government's debt.

Perhaps the most important thing that Ottawa does with your tax dollars is to transfer money to persons (including you) and to other levels of government (mostly to the provinces).

Let's look at transfers to persons. From 1982/1983 to 1993/1994, for nine of these 12 years, these transfers were about 5 percent of GDP. In four of these fiscal years, the transfers to persons amounted to from 5.7 percent to 5.9 percent of GDP. However, after Paul Martin became Finance minister, they fell like a big rock.

Some people will say, "So what? Martin fixed the long string of Liberal and Conservative deficits." Others, however, will say that the deficits could have been much more justly addressed with changes to the tax system. You have already read about this in the tax section of this book. While for years transfers to persons were over 5 percent of GDP, in most recent years they've been in the 3.8 to 4 percent range.[5] The last time they were as low as 3.8 percent was 38 years ago, way back in 1970/1971.

Now let's look at Ottawa's transfers to other levels of government. As recently as 1983/1984, these amounted to 4.2 percent of GDP. But by 2000/2001, they were down to only 2.3 percent of GDP, a decline of $23.4-billion, although they increased to 2.9 percent in 2006/2007. In the Harper government years, transfers to other governments have been and are planned to be only 2.8 or 2.9 percent of GDP through to 2012/2013.

The impact of these massive reductions in transfers has inevitably fallen, and will fall, mostly on the poor and on those in the lower- to mid-income levels, as you have already seen in the chapters on poverty, social programs, welfare, and distribution of income.

As I have said, the two largest components of Ottawa's spending are transfers to persons ($69.7-billion in 2006) and transfers to provincial and local governments ($52.7-billion in 2006). In 2005, individuals on average received $15.54 in government transfers for every $100 of employment income.

The next time you encounter some anti-tax, anti-government slasher from the raging right, ask them which of the above two categories of transfers they would like to see chopped, and then ask them if they would consider running for public office featuring such a policy in their campaign. Ask them for which of the following would they like to cut spending: health, education, the environment, child benefits, pensions, or defence.

While the provincial premiers constantly clamour for more money from the federal government, our cities are increasingly shortchanged by the provinces. *Globe and Mail* columnist Neil Reynolds has pointed out that while the revenue of municipalities is about 4.4 percent of GDP, they spend about 6.95 percent, while relying on the provinces for some 40 percent of their revenue.[6]

Again, how is Canada doing in relation to our central government's debt and debt charges? There was a time, not too long ago, when every public opinion poll showed that these two items were near the top of the list of public concerns. But this has all changed. Back in fiscal year 1995/1996, Ottawa's public debt charges were $49.4-billion. By 2006/2007, these were down to $33.9-billion, having declined for five years in a row, down from a high of 6.6 percent to 2.3 percent of GDP in 2006/2007. They are expected to fall further, to 2.1 percent of GDP in 2008/2009, or some $33.7-billion. Bear in mind that, in 2006, only 1 percent of GDP in Canada was some $14.4-billion. Of course, some of our reduced annual debt payments has gone to help fund health care, education, child benefits, and so on.

For decades, some people have worried, as they should, about the federal government's debt and the money leaving the country to pay the large interest on it. At one time, non-residents held almost 28 percent of the debt. But today that foreign debt is down to only 13.1 percent and headed lower, a dramatic decline.

Some people (and some intentionally) confuse the gross debt and the net debt. The net debt is the gross debt minus government assets. As recently as 1999/2000, the gross debt was just under $716-billion, and the net debt was $576.3-billion. By 2006/2007, the net debt was down to $467.3-billion, and it is expected to be down further in 2008/2009.

As a percentage of all of Ottawa's spending, debt charges have fallen from a high of 29.8 percent in 1996/1997 to only 15.3 percent in 2006/2007.

As indicated earlier, most Canadians place health care, the environment, education, and child poverty at the top of their list of concerns. Yet an October 2006 *National Post* poll shows that Canada's business leaders want to allocate government surpluses to paying down the debt. Corporate executives are three times more likely than other Canadians to support debt reduction, and five times more likely to support tax cuts. But when this *National Post* poll was taken, government's net financial liabilities as a percentage of GDP were already at their lowest level in a quarter of a century and headed lower – and as we have seen, they were by far the lowest of any of the G8 countries.

In 2007, the OECD projected that Canada's general government debt interest payments would be a lower ratio to GDP than in 18 other OECD countries and well below the OECD and the Euro area averages.

Moreover, most economic projections indicate that even with no further paydowns of the federal debt, it will continue to shrink substantially as both a percentage of GDP and as a percentage of all government spending.

One interesting but sad development in government revenues in Canada is the rapid rise in gambling revenues. In 1992, these amounted to $2.73-billion. By 2006, they had ballooned to $13.3-billion.

Now, for the benefit of Thomas d'Aquino of the Canadian Council of Chief Executives and his continentalist colleagues who are so determined to adopt American standards, values, and policies, and are so gung-ho for greater Canadian harmonization with and integration into the United States, let's look briefly at the debt situation there.

When George W. Bush became president, the U.S. federal debt was $5.7-trillion. By the end of 2006, it was an astonishing $8.5-trillion. The head of the U.S. Government Accountability Office in Washington has warned that unless major corrective steps are taken, the U.S. debt will increase by many trillions of dollars in the foreseeable future.

Nevertheless, despite such warnings, more than a few reputable economists are forecasting U.S. deficits of hundreds of billions of dollars a year for as far as they can see into the future, and even worse trade deficits.

By the summer of 2006, the foreign holdings of U.S. government debt were increasing rapidly. U.S. government interest payments to foreigners long ago passed all government spending on education, social services, and job training combined.

About $2.2-trillion (U.S.) of the American public debt is owned by China, Japan, and the OPEC countries. By the time George W. Bush leaves office, the national debt is projected to be $9.4-trillion.

Lester R. Brown of the Earth Policy Institute writes:

> China is now becoming our banker. This developing country, where income levels are one sixth those in the United States, is financing the excesses of an affluent industrial society. What's wrong with this picture? . . .
>
> If China's leaders ever become convinced that the dollar is headed continuously downward and they decide to dump their dollar holdings, the dollar could collapse.[7]

As *The New Yorker* put it,

> More than any other nation in history, the United States depends economically, on the kindness of strangers. Right now, Asian investors are very kind. Asian central banks already own trillions of dollars in American assets.
>
> Of course, the Chinese and the Japanese could decide that costs of the falling dollar are too great, and suddenly stop (or, at least cut back sharply) their lending to the United States.[8]

And, if they did this, what would be the result? For certain, it would be a very "hard landing," with terribly high interest rates, high inflation, a plunging stock market, and even more depressed house prices. Not just a recession, but possibly a prolonged and serious deep depression.

Meanwhile, in the 10 years from 1996 to 2006, the U.S. current account deficit grew from less than 2 percent to almost 7 percent of GDP, while in 2007 Washington's debt stood at just under 37 percent of GDP.

As well, the American trade deficit in 2005 was $725.8-billion (U.S.), and it was $763-billion (U.S.) in 2006, the fifth consecutive record trade deficit. The last U.S. trade surplus was 32 years ago, in 1975.

In 2006, there was a fifth consecutive record combined current account and fiscal deficit. In January 2006, the president of the Federal Reserve Bank of New York, Timothy Geithner, warned that the soaring deficits were not only a significant threat to the United States, but also a threat to the global economy that could not be sustained.

Please keep these U.S. figures in mind when you read in the conclusion of this book how Mr. d'Aquino and friends want to integrate us even further into the United States.

One last point about Ottawa's transfers to the provinces for social programs. As the Canadian Union of Public Employees has pointed out in their comments on the Harper government's 2007 budget, if these transfers were only set to keep pace with the increase in population and inflation, they would be a huge $4.4-billion higher in 2008/2009. Imagine what some of that money could do for child poverty, foreign aid, social housing, post-secondary education, or cancer research.

And one last reminder about total Canadian government revenue and spending as percentages of GDP compared to the other 29 OECD countries: In revenues we're in 19th place, in spending we're even lower, down in 20th place.

Now, please go back and look at the Stephen Harper quote that opens this chapter.

DECENTRALIZATION

MEECH LAKE AND CHARLOTTETOWN, WHETHER YOU LIKE IT OR NOT — A GUARANTEED RECIPE FOR DISASTER

I s Canada, as so many claim, a badly over-centralized country?

Not so. Follow the money. All of the following 22 OECD countries have a higher share of all tax revenue going to their central government than Canada: Australia, Austria, Switzerland, the Czech Republic, Denmark, Finland, Greece, Hungary, Iceland, Ireland, Italy, Korea, Luxembourg, the Netherlands, New Zealand, Norway, Poland, Portugal, the Slovak Republic, Sweden, Turkey, and the United Kingdom. Only six OECD countries had a lower share going to the central government.

Beginning in the 1960s, and in every decade thereafter, the federal government's share of all government income has been shrinking. Where Ottawa was once receiving over 65 percent of all government revenue in Canada, by 2006 that was down to 39.2 percent (well below the OECD average of 49.4 percent for federal central governments). Today, in a list of the top 50 developed countries, 34 central governments do a larger share of all government spending than Ottawa.

In his *Globe and Mail* column on November 10, 2006, Jeffrey Simpson referred to the "decentralized political system" in the United States. "Decentralized" in some ways, true, but certainly not in government spending and in the control by Washington of huge areas such as resources, education, and social policy. In 2005, Washington was responsible for 64 percent of all government expenditures in the United States, and only 9 percent of this was in the form of federal grants to state and local governments. As percentages of GDP, Washington spent 20.4 percent in 2005, while state and local governments combined spent only 13.5 percent.

The decentralization of Canada has been progressing rapidly since the days of Brian Mulroney. And it's clear that Mr. Harper, if he remains in office, will not only continue this process, but accelerate it. This despite the fact that many thoughtful and knowledgeable Canadians believe Canada is already so decentralized it is close to being ungovernable.

In 2006, looking at total government spending, provincial expendiures were 47.8 percent of the total, and local government spending was 16.7 percent, leaving Ottawa with only 35.5 percent.[1]

Bear in mind, Ottawa's transfers to the provinces and local governments in 2006 came to almost $56-billion, while provincial transfers to local governments were $39.4-billion. Subtracting these two numbers, the provinces and local governments spent over twice as much as Ottawa's total.

How does that compare with the ignorant claims from Alberta and Quebec, and from some our right-wing business leaders and newspaper columnists, about Canada being "over-centralized"?

As for Ottawa's ability to introduce new pan-Canadian programs, Tom Walkom of the *Toronto Star* has it right:

> In February 1999, Chrétien's Liberal government signed a deal with all provinces except Quebec in which it promised not to introduce any more programs in areas of provincial jurisdiction (such as health and education) unless a majority of provinces agreed.
>
> This so-called social union framework agreement did not lead the television newscasts the next day. Nor did it give rise to "the world is ending" headlines in the daily press.
>
> But, by limiting Ottawa's capacity to create new pan-Canadian social programs, it carried far more real weight than Harper's Quebecers-are-a-nation resolution.

If Chrétien's social union framework agreement had been in place in 1968, for instance, it is not at all clear that Ottawa would have set up a national medicare system. At the time, only two provinces were unequivocally willing to join up.[2]

Subsequently, the Paul Martin government negotiated a list of one-off deals with the provinces. Now Stephen Harper is intent on entrenching shared-cost programs, allowing any province to opt out, yet receive full compensation and proceed to set their own rules and standards. As *Globe* columnist John Ibbitson has written (January 2006), this and other plans Ottawa has to gain favour in Quebec are "the essence of the Meech Lake and Charlottetown accords, the failed social union negotiations."

Jeffrey Simpson says that "open federalism" is the order of the day in Ottawa: "It began under Prime Minister Paul Martin who cut so many one-off deals with the provinces that his style became known as 'delicatessen federalism.' Open federalism is essentially a vision of Canada in which no national objectives are established that have not previously been vetted and okayed by the provinces."[3]

And then, Simpson continues, came Stephen Harper: "This Prime Minister argued that people should build their own 'nations' or provinces, even to the point of 'autonomy.' He is implicitly, therefore, rejecting any idea that Canada might be something more than the sum of its parts."[4]

The *Toronto Star*'s Carol Goar asks, "Who loses?"

First, anyone who would rather have strong national safety nets.

Second, anyone who believes that 32 million Canadians can achieve more collectively.

It's an acceleration of the devolution that began under the Liberals. They withdrew from urban affairs, housing, consumer protection and job training. They shrank unemployment insurance. They replaced public servants with contractors and consultants.[5]

But do we really want a country where the government is bound and shackled, down on its knees before powerful, parochial, provincial potentates?

Of course, some of the provincial premiers are delighted by all of this, especially by any new tax point transfers that will forever be completely irreversible. An already comparatively weak federal government could be turned into one that is completely dysfunctional, a government where provincial and territorial spending increasingly exceeds that of the national government, in a country with declining common standards and goals. And of course the premiers will never be satisfied. Enough will never ever be enough.

All this – combined with the onerous clauses in NAFTA already described that make Ottawa vulnerable to American government-supported corporate legal action – has helped produce a country where new national programs of consequence are increasingly impossible. Clearly, a weak central government combined with an excessive and growing level of foreign ownership and control is a guaranteed recipe for disaster. Yet, somehow, most of our political commentators and politicians seem to be completely blind to the potentially catastrophic consequences.

The combination of Stephen Harper and a separatist Quebec government would certainly mean a major transfer of even more power to Quebec. And anyone who thinks Alberta will not immediately demand every single thing received by Quebec is hopelessly naive. Then, of course, will come similar demands from British Columbia and Ontario.

Welcome to the new balkanized Canada, ready to be picked off, one region at a time, by American oil, natural gas, electricity, water, and other resource corporations. Canadians debated and rejected Meech Lake and the Charlottetown Accord. But now, thanks to our spineless politicians, we're headed for a chopped up, colonial status.

Bob Rae, writing in the *Globe and Mail,* put it this way:

> For the first time in our history, we have a prime minister and a cabinet talking openly about giving up the game for the federal government. . . .

What could be so bad about a federal government's agreeing not to spend money in "areas of exclusive provincial jurisdiction"? Looking backward and forward, we can find a simple answer: a lot. . . .

Whether introduced by Conservatives (progressive or otherwise) or Liberals (in majority or minority governments), our history as a country has been marked by moments of a pan-Canadian vision led by federal governments with the support of Parliament and people.

This is what Mr. Harper wants to end, either by constitutional amendment or federal-provincial agreement. For the first time, we have a national party – the Harper Conservatives – ideologically committed to a fundamentalist misreading of our history and Constitution, and a separatist party only too happy to reduce the Canadian government to a marginal role, just before, in their sad dreams, it disappears altogether.

Canadians who want their federal government to support early childhood education, decent housing, cities that work, a healthy environment, new initiatives in health care, more mobility for students, better research and stronger universities should be appalled at this emasculation.

Thirteen fiefs putting up more walls and moats between themselves and their neighbours does not make a country.[6]

In an editorial, the *Toronto Star* described Stephen Harper's recognition of Quebec as "a nation" as "a reckless move." Now, according to the editorial, Quebec Premier Jean Charest expects Ottawa "to formally surrender the ability to create and support nationwide programs like health care, education, pharmacare, child care, welfare and manpower training. That is what Quebec nationalists have long demanded."[7]

And how do Canadian feel about all of this? When asked if the federal government should be involved in areas of provincial jurisdiction, 67 percent agreed and only 29 percent disagreed. Once again, as in the case of foreign ownership and control, our hopelessly myopic politicians are leading us in exactly the opposite direction from the will of the people.

And it is a fatal direction.

But what would happen if Stéphane Dion ever becomes prime minister? Thomas Walkom's answer leaves little doubt. The columnist describes an occasion when, in the presence of Dion, "someone raised the social union framework agreement, the 1999 federal-provincial pact designed to limit Ottawa's spending in areas of provincial jurisdiction such as health. At the mention of the agreement, the Liberal leader's eyes lit up. He extolled its virtues; he explained at some length why it was such a good idea."[8]

Stéphane Dion has supported Harper's Quebec-as-a-nation proclamation. Before that, Dion supported the failed decentralizing Meech Lake accord. And the NDP? They are willing to give Quebec total authority to opt out of federal social programs, but the province would get to keep the taxpayers' cash anyway.

Poor Canada. Poor Canadians. What awful political leadership we have. What a tragedy that the leaders of our two major political parties are prepared to turn us into a hobbled, fractionalized, enfeebled country and the NDP seems quite willing to go along with the process.

ENERGY POLICY IN CANADA

FROM THE RIDICULOUS TO THE ABSURD

"The dumbest people in the oil patch"

H ere, in no particular order, are some of the things Canadians should find incomprehensible:

- Royalties in Alberta's oil sands have for many years probably been the lowest in the world. Critical witnesses attending public hearings on petroleum royalties and taxes in Alberta in 2007 called the remarkably profitable energy industry "the most heavily subsidized business in Alberta" and said that "foreigners look at us as if we might be the dumbest people in the [global] oil patch."[1]

- Overall petroleum taxes, royalties, and revenue-sharing in the petroleum industry in Canada have been less than a third of those in Norway.

- We're quickly using up our natural gas reserves. In the words of Statistics Canada, "The need to find new gas reserves is relentless." Natural gas production will start declining in 2009, falling by as much as 15 percent.

- We now have less than nine years of natural gas left, and less than 10 years of proven conventional oil reserves left. New untapped reserves are more difficult to find, and recent discoveries are smaller and more difficult to produce.

- Despite this, we're exporting massive and increasing quantities of oil and natural gas out of the country to the United States, and allowing the petroleum companies to plan and build huge new export pipelines. In its 2007 budget, the Harper government introduced new tax concessions for new natural gas pipelines.
- Despite all the terrible problems relating to oil sands production (greenhouse gas production, water use, natural gas requirements), we're planning massive increases in the number of new plants and production.
- Despite their enormous profits, oil companies are still subsidized and we provide them with billions of dollars in tax breaks every year. In 2000, the primary oil and gas industry in Canada recorded an all-time record operating profit of $23.8-billion. In 2006, the industry's profit was $39.9-billion.
- No other country in the world would have been stupid enough to have agreed to the mandatory sharing and ridiculous pricing provisions of NAFTA.
- No other country would have sold off the ownership and control of so much of its petroleum resources.
- Every country you can think of has well-planned energy security policies in place. Canada has no such thing. Canada is one of only two of the 26 members of the International Energy Agency not to have a strategic petroleum reserve. On the contrary, our policy is to ship all of our oil and natural gas out of the country as quickly as we can.
- Gordon Laxer, of the University of Alberta and the Edmonton-based Parkland Institute, points out that "the United States has a National Energy Policy (an NEP) that emphasizes self-sufficiency, energy independence and domestic ownership. Canada has an NEP, only it stands for No Energy Plan."[2]
- In a country that should be the envy of the world for its current and potential hydro resources, we still have no east-west electricity power grid.

- Many countries consider planning their resource use with the needs of future generations in mind. Apparently, no one in government in Canada believes there will even be future generations.

- Where at one time reserves of natural gas were held back for future Canadian consumption, thanks to Brian Mulroney and Jean Chrétien and the completely absurd sections of NAFTA dealing with energy, the border between Canada and the United States has disappeared. By the time the Mulroney government was elected, Americans had begun to regard Canadian oil and natural gas as essentially their own. Where once it was required that a 20-year reserve for Canadian use was mandatory before any approval of exports, now the petroleum companies can ship the gas out of the country just as quickly as their pipelines can be built.

- In Stephen Harper's "world energy superpower" of Canada, about 40 percent of the oil used in Ontario is imported, as is almost 90 percent of the oil used in Quebec and the Atlantic provinces (almost half of it from reliable places like Iran, Algeria, and Saudi Arabia). No one in government appears concerned about the inevitable future disruptions of supply.

- Any kind of sensible energy security policy would have Alberta oil and natural gas flowing to Central and Eastern Canada instead of being exported to the United States, and Atlantic natural gas would be used in the Maritimes and Quebec instead of being exported to the United States.

- Even in Alberta, public opinion polls show that such policies would meet with strong approval.

Okay, let's expand a bit on some of these items. In 1988, the year before the FTA came into effect, Canadian natural gas exports amounted to 36.25 million cubic metres. By the time NAFTA came into effect, they were up to 72.36 million cubic metres, and by 2005 they were almost three times the 1988 rate, having climbed to 104.84 million cubic metres.[3]

The following are among those who are delighted: George W. Bush, Dick Cheney, Exxon, EnCana, and the Alberta government.

In 2006, the oil sands accounted for 46 percent of all crude oil produced in Canada. By 2020, only about 20 percent of Canadian oil production will come from conventional sources, the remaining 80 percent will come from the oil sands.

In 2006, production of domestic crude oil and equivalent hydrocarbons in Canada was at an all-time high, "thanks," in Statistics Canada's words, "to strength in Alberta's oil sands." Production was a record 152.9 million cubic metres (one cubic metre is the equivalent of 6.3 barrels). The latest estimates are that our oil exports will more than double by 2015, and will more than quadruple by 2025. By 2025, conventional oil production will have fallen by almost 60 percent.

Exports of natural gas took 59.4 percent of all Canadian marketable gas production in 2006, while exports of crude oil and equivalent hydrocarbons were 67.6 percent of production.[4]

How does all of this compare with future demand?

Some forecasts suggest that by 2008 only minimum growth in oil demand will be met, and by 2010 there will be no growth at all.[5] Around the world, demand is growing. The United States, with 300 million people, now consumes 25 percent of all oil used annually, but China, with 1.3 billion, uses only 9 percent. This, however, is changing very rapidly. China has already replaced Japan as the world's second largest oil importer after the United States and is putting an average of 14,000 new cars on the road each and every day.

Is the world running short of oil? Some experts say worldwide production peaked in 2006. North Sea production is in a steep and irreversible decline, and the United Kingdom has found it necessary to import oil for the first time in decades. Production from Mexico's largest field is in decline. Others say the peak for oil extraction will come two years from now, in 2010. Meanwhile, there will continue to be huge increases in demand, a demand that can't possibly be met by new discoveries. Some analysts are predicting oil prices in the near future as high as $150 a barrel.

In November 2007, the International Energy Agency (IEA) predicted that an "alarming" increase in energy demand will threaten global energy security and speed up climate change. In "the most pessimistic overview of the world energy market we have ever portrayed," the IEA predicts an unprecedented growth in demand as the world's overall energy needs rise "well over 50% higher in 2030 than today while demand from China and India alone is expected to double in the next two decades." By 2030, India and China together will import as much oil as the United States and Japan do today.[6] Or, try to import.

Some estimates suggest that in the not-too-distant future up to 80 percent of the world's future conventional oil supply will have to come from the Middle East. The International Energy Agency says while demand for crude is increasing very substantially, supplies from non-OPEC countries are declining.

Are there those who *don't* believe that global oil discoveries have been in precipitous decline? If there are, they must be living deep in a cave. Every decade since the 1950s, new discoveries have fallen sharply, from a high of 470 billion barrels in the 1950s, to only 120 billion barrels in the 1990s. Two studies, one by the U.S. National Petroleum Council and the other by the IEA, both forecast that it will be impossible to meet soaring demand, that a supply squeeze will occur as early as 2012, and a severe crisis will affect all major petroleum-consuming countries by 2030 at the latest.

Listening to the nonsensical industry propaganda from the "quickly, we must have more exports" producers in Calgary, you would think there's no reason in the world for concern. That's because they include "waiting to be discovered" figures in their reserves. But actual reserve figures show a steady decline.

Since Ottawa's castration of the National Energy Board, oil and gas exports have been largely determined by the producing companies, now majority foreign-owned and controlled. Is there anyone anywhere who believes for a moment that these foreign firms have Canada's best interests in mind, or that they might ever consider responsibilities to future generations of Canadians? Mind you, few Canadian-owned petroleum companies are any better.

As for the National Energy Board, in an April 2007 letter to Gordon Laxer, they say that "unfortunately the NEB has not undertaken any studies on security of supply." Surely an astonishing admission of government incompetence in a country with such increasing exports and diminishing reserves.

Meanwhile, Enbridge and ExxonMobil are working on a major pipeline to supply oil to Illinois and Texas that could move 400,000 barrels a day. TransCanada is proceeding with a new pipeline to southern Illinois, with links to Oklahoma and Texas. Altex Energy is working on a $3-billion, 250,000-barrel-a-day pipeline to Texas. An ExxonMobil pipeline from Canada is now delivering mostly raw bitumen to Louisiana, and Kinder Morgan is planning yet another pipeline to Texas.

But fear not, a July 20, 2007, heading in the *Globe and Mail* tells us that, according to a U.S. energy economist, "There's No Downside for Canada" from all the new pipelines and vastly increased exports.

So is there a chance any of this is going to change? Not according to the U.S. Energy Administration in Washington, as reported by CanWest News Service:

> Canada – which in 2005 replaced Saudi Arabia as the single-largest supplier of energy to the U.S. – will continue that position over at least the next two decades, thanks to the multi-billion oil sands developments in Alberta.
>
> Canadian exports to the U.S. will reach 2.6 million barrels per day by 2030, compared with current levels of just over one million bpd.[7]

In October 2006, a new Natural Resources Canada study[8] predicted declining natural gas production and increasing reliance on expensive, polluting oil sands production. Shawn McCarthy of the *Globe and Mail* put the document in the correct perspective: "The report represents a stark contrast to the message of bullish federal and provincial politicians and oil industry officials who have assured consumers in the United States that they can rely on Canada to help meet their growing energy needs."[9]

The study says Canada's natural gas production will peak at 6.6 trillion cubic feet a year by 2011, and then it will decrease. While coal-bed methane and Mackenzie Delta natural gas may help soften the impending decline in gas supply, they will come nowhere near reversing the sharply downward trend expected to develop within the next three to four years.

On the subject of our natural gas exports, the *Globe and Mail*'s Eric Reguly writes: "Imagine the premier of Ontario explaining to voters that the province can't build a Kyoto-friendly natural gas plant to replace the doomed coal burners because Alberta needs the clean fuel in Fort McMurray to make oil for American SUVs."[10]

Even if you don't hear much (or anything) about it from either Calgary or Ottawa, even the Americans now realize that Canada is running out of natural gas. Yet despite the fact that our reserves peaked in 2001 or 2002, we're still mindlessly sending most of our production to the United States. By the spring of 2006, the average initial production rates of the best new gas wells was already down almost 25 percent from the 2002 rate. The latest U.S. estimates are that Canada is going to run short of natural gas despite "heroic" industry exploration efforts.

One of Canada's leading government energy experts from Calgary, who asked not to be identified by name, described the natural gas situation to me as "pretty scary."[11]

There are some who have been led to believe that Canada's reserves of Arctic natural gas will help alleviate concern about our dwindling supplies. Not so. If and when the planned pipeline is ever built, much of the Arctic gas will go straight to Northern Alberta for use in the oil sands. Just one Cold Lake oil sands project consumes 100 million to 150 million cubic feet of natural gas a day. Producing a barrel of tar sands oil requires from 1,000 to 2,000 cubic feet of natural gas, plus between four and six barrels of water. Just think about that for a moment. It's becoming clear that a limited natural gas supply may be a crucial problem for much greater oil sands production. And estimates suggest that new plants are going to need an extra 180 million litres of water a day, over and above the 140 million now committed. There is already great concern that water levels in the Athabasca River are more than 30 percent lower than they were in 1970:

> The amount of water available in Northern Alberta isn't sufficient to accommodate both the needs of burgeoning oil sands development and preserve the Athabasca River contends a study issued jointly by the University of Toronto and the University of Alberta.
>
> "Projected bitumen extraction in the oil sands will require too much water to sustain the river and Athabasca Delta, especially with the effects of predicted climate warming."[12]

Should we be exporting our fast-shrinking natural gas reserves so rapidly? Even Jim Dinning, former Alberta Conservative provincial treasurer, has suggested that Alberta should have a 50-year supply of natural gas before more is exported from the province. Dinning provides the following great quote: "Injecting natural gas into the oil sands to produce oil is like turning gold into lead."

By the summer of 2006 we were producing about 2.5 million barrels of oil a day. Some estimates suggest that by 2015 this will increase to 4.5 million barrels a day, of which over three million will come from the oil sands. In 2006, we were exporting over 1.7 million barrels a day to the United States and at the same time importing some 750,000 barrels for refineries in Eastern Canada.

On the surface, if bullish estimates are reasonably accurate, all might seem well. But as we will see, given that the Alberta oil sands are the biggest contributor to the growth of greenhouse gas emissions in Canada,[13] and given that Alberta, with only just over a quarter of Ontario's population, already emits more greenhouse gases, and given that oil sands production is scheduled to triple within the next decade, and that some 75 percent of the production is for export to the United States, and that the Alberta economy is already badly overheated – well, given all that, what do you think we should be doing?

Alberta Finance Minister Lyle Oberg warns Ottawa not to touch the expensive special tax concessions for the oil sands, which amount to a subsidy of over $425-million a year for the petroleum companies. You know, those poor petroleum companies whose remarkable records show that, after royalties, after tax, after depreciation and all other expenses, they made net profits of $85.7-billion from 2003 to 2005 inclusive and still somehow managed to pay an after-all-expenses tax rate of only 29 percent.

For amusement, try this. At royalty hearings in Alberta, the president of Canadian Natural Resources warned that higher royalties would mean "drastically" lower energy activity in Alberta: "It's a myth out there that this is a hugely profitable business."[14] Perhaps there's someone out there concerned about poor Imperial Oil, majority-owned and firmly controlled by Exxon Mobil Corp., whose 2006 profit of $39.5-billion was the largest in U.S. corporate history and whose market capitalization has exploded from $80-billion to $360-billion. Or perhaps they might be concerned about Chevron, the second-largest U.S. oil company, with three consecutive years of record profits, $17.1-billion in 2006. Poor Imperial Oil saw its profits increase in 2006 by only 15 percent over 2005, to $3.04-billion, while EnCana's profits were expected to be $7.5-billion, the biggest-ever in Canada's corporate history.

In a June 2007 study, Shawn McCarthy, writing about comments by the leading Edinburgh-based consulting firm Wood Mackenzie, said,

> World oil companies have hit a gusher in Canadian tax policy. In the past five years, Canada is the only significant oil and gas producing country to actually reduce its share of oil revenues.
>
> Many other oil-rich jurisdictions, including Britain and Alaska, have significantly increased their share of the revenue generated by rapidly rising prices. But, as a result of cuts to federal tax rates introduced by the former Liberal government, oil companies have seen their tax bite reduced. Tax rates in Canada are lower than they were when prices started to rise.[15]

Only in Canada, you say?

Oh, and by the way, according to Albert Koehl, a lawyer with the Sierra Legal Defence Fund, "the federal government has handed out more than $2 in tax subsidies to these companies for every $1 it spent on Kyoto compliance."[16]

Around the world, state oil companies now control between 80 and 90 percent of the world's oil and gas. ExxonMobil, the world's most valuable company, is only 14th in the world in reserves, after 13 national oil companies.[17] As John Warnock pointed out in a paper for the Parkland Institute, the huge profits accompanying the petroleum price increases have largely gone to the OPEC country treasuries. But most of the windfall profits in Canada and the United States have gone to large private corporations, many of them effectively able to siphon the profits off to tax havens by selling at a low price to a subsidiary in one of the tax havens and then selling it into the market at a much higher price. According to Warnock, by 2003, "58% of U.S. corporate profits were taken in offshore tax havens."[18]

Returning briefly to the question of who should own our oil and natural gas, in a September 2006 poll, when asked, "Do you think provincial resources should belong to all Canadians?" 76 percent said yes and only 21 percent said they should belong to the people who lived in the province where they were produced. Even in Alberta, 55 percent said resources should belong to all Canadians, while only 39 percent said only to the people who lived in the province.[19] Yet by 2004, 55.2 percent of all oil and gas extraction revenue was already going to foreign-controlled firms. The figure for 2007 isn't out yet, but you can bet it will be close to or above 60 percent.

By the way, roughly three of every four Canadians have been against the federal government selling its share in Petro-Canada.

As for what Albertans and Canadians receive from the exploitation of the oil sands, it's difficult to believe our politicians could really have been so incredibly stupid. Writing about how Alberta lowered its oil sands royalties to 1 percent until producers recovered their capital costs, the *Globe*'s Eric Reguly said,

The change amounted to one of the biggest energy giveaways ever . . . one of the lowest oil royalty rates on the planet. . . . Between 1995 and 2002 Norway collected an average of $14.10 (U.S.) a barrel in royalties, compared with just $4.30 (U.S.) for Alberta, according to a report by the Pembina Institute: "And then Ottawa chipped in with an accelerated capital cost allowance, allowing write-offs of 100 percent against first year income."

Reguly asks, "Why should taxpayers have to subsidize the world's most profitable product?"[20] Incredibly, even as oil prices dramatically escalated to record levels, Ottawa's take headed downhill. Federal corporate petroleum industry taxes brought in $5.1-billion in 2005, but these are expected to plummet to $2.4-billion in 2008, even though the industry will take in a massive $100-billion revenue from 2006 to 2008.

At the same time, Alberta has been facing a large drop in royalty revenues despite rising production and high prices. In 2006, oil sands royalties were $2.3-billion. By 2009/2010, they were expected to be only $1.1-billion, the same level they were five years earlier on much smaller production levels. Total petroleum royalties were expected to fall from $9.8-billion in 2006/2007 all the way down to $6.6-billion.

In November 2006, the Pembina Institute said that Alberta was receiving almost a third less in royalties from a barrel of oil than it did 10 years earlier. It said, "In short, citizens are losing out, while corporations are winning." And according to consulting firm Wood Mackenzie, "The feverish pace of development that the tax and royalty regime are facilitating is resulting in significant negative environmental and social impacts."

While the Alberta government has moved at the pace of a turgid snail, *The Economist* tells us that "all around the world, from Algeria to China, governments are changing the terms of investment in oil and gas on the grounds that they are not receiving their fair share of profits."[21]

Remarkably, Suncor Energy was projecting that its oil sands royalties could fall to as low as 6 percent in the years 2009 to 2012. In 2006, Suncor recorded a profit of $3-billion. Despite huge increases in oil sands production, Alberta Department of Energy predictions were that oil sands royalty revenues will be the same in 2020 as they were back in 2004/2005.[22]

It truly boggles the mind.

A bit more about the incredibly stupid energy provisions in Article 605 of NAFTA. Since the agreement came into effect, oil and natural gas exports have increased dramatically. Can we cut these back if we deem it to be in our own national interest to do so? Sure we can. But only if at the same time we cut back our own consumption. NAFTA says that Canada must continue exporting the same proportion of its oil and gas production to the United States as it did in the previous three years. (The proportionality clause refers also to petrochemical goods and hydro power.)

Gordon Laxer discussed energy and NAFTA in a *Globe and Mail* piece:

> Our NAFTA partners are looking after their own energy security. Canada is the only NAFTA country prevented from doing so.
>
> Four years ago, the United States adopted a national energy policy that emphasizes energy security, self-sufficiency, and even financial support for domestically owned firms. The terms are reminiscent of Canada's national energy program in Pierre Trudeau's time.
>
> When NAFTA was being negotiated in 1993, oil and gas corporations based in Canada, many of them foreign-owned, lobbied for a proportionality clause to be included in the agreement.[23]

How did the Mexicans, our other NAFTA partner, react to the petroleum industry proportionality proposal strongly backed by the U.S. government? They scoffed at the idea, and quickly and firmly rejected it. So we ended up with a national energy policy for the United States and a ridiculous, locked-in, straitjacketing, useless no-energy-policy for Canada.

Then there's the question of price.

Want to ship oil or natural gas to a purchaser who'll pay a much higher price than the United States? To China, for example? Forget it. We can't cut back the proportion of production that we sell to the United States. And NAFTA says we can't charge Americans a penny more than we charge Canadians.

Has *anyone* in Canada stopped to think for even a moment what would happen if there was a major crisis in the Middle East and oil supplies to the United States were cut off? We know for sure that prices in the United States would go up sharply. And because of NAFTA, Canadian prices would do the same.

So, all this considered, what should we do? You already know the answer if you've read the chapter on NAFTA. I believe we should abrogate NAFTA and impose export taxes on all oil and gas exports. We should use the resulting tax revenue to fund R&D and the production of renewable energy. The Americans, and the oil companies, and the Alberta government will scream blue murder. But did the Americans stop importing oil when it spiralled to over $98 a barrel and their deficit with OPEC jumped from $10-billion in 1988 to $80-billion in 2005? Are the oil companies suffering? Is the Alberta government short of money?

As for other things that need to be done, here are some suggestions from the excellent Pembina Institute and from other experts:

- We should give government subsidies to promote energy efficiency, renewables, and conservation instead of subsidies to oil companies.
- We should plan a much more moderate pace of development of all forms of petroleum production.
- We should ensure that environmental destruction by the petroleum industry is properly penalized, and that the revenue is used to improve the environment and for R&D on renewables.
- We should take steps to improve and expand public transportation vastly, and promote and legislate much more mandatory car pooling.

- We should impose lower taxes on smaller, more fuel-efficient, less polluting cars, and bigger taxes on the rest, with rebates for low-income individuals and families.
- We should legislate much tougher auto efficiency standards and vastly increase rail transport.
- We should study Brazil's great success with ethanol and decide how we can best develop our own cellulosic fuel-efficient ethanol.
- We should increase taxes on petroleum company profits.
- We should stop exporting raw bitumen and instead process it into upgraded products here in Canada, making sure that the carbon produced is properly captured. Even former Alberta Premier Peter Lougheed says that it's unacceptable that we should be shipping jobs down the pipeline.
- We should dedicate ourselves to becoming a world leader in the production of improved new wind turbines and solar panels.

In 2004, Germany was producing 40 percent of all wind power, Spain 20 percent, the United States 16 percent, Denmark 7 percent, Italy and the Netherlands 3 percent each, Japan and the United Kingdom 2 percent each, and all others combined 7 percent. Today, Canada stands in 13th place in the production of wind power. By 2005, Denmark was getting 20 percent of its electricity from wind and Germany over 7 percent. In Canada, in 2005, wind, solar, and tidal power combined made up only 0.5 percent of electricity generation. Yet a 2005 poll said over 90 percent of Canadians are in favour of more investment in solar, wind, and hydro development, while 76 percent were opposed to more coal production and 64 percent were against increased nuclear capacity.[24] But that's exactly what Stephen Harper and Ontario Premier Dalton McGuinty are planning.

The good news is that in 2006, wind farm construction in Canada doubled from 2005, producing enough energy to power some 370,000 homes. Some recent estimates suggest that by 2009 we could be among the top five wind-power producers in the world, with close to $18-billion in investments planned between 2007 and 2015. Other estimates suggest that up to 5.5 percent of Canada's energy will come from wind power by 2015.[25] Unfortunately, because of our failure to do our own R&D, we have to buy almost all our wind-power turbines from American or Danish firms.

When you read the following quote, consider if you have heard anything remotely similar from our prime minister, the premier of Alberta, or any of their cabinet ministers. Here, from October 2006, are the words of Norwegian Secretary of State Geir Axelsen, who was speaking about Norway's oil:

> When we have this natural resource, it's a moral question: Is it right for our generation to just spend this wealth in our time? It's taken 100 million years to create these resources, and it's possible that they will be used up in the next 50. Isn't it reasonable that the next generation also has the option to choose how to spend the money?

The Norwegian petroleum company Statoil ASA, 70 percent owned by Norway, plans a multi-billion-dollar, 200,000-barrel-a-day oil sands project in Alberta, but says that new environmental costs will not be a problem; the company, which operates in 35 countries, is already a leader in carbon capture and storage.

This said, the previously discussed natural gas and water shortages are clearly going to be a very difficult problems, perhaps impossible to solve. Increasingly, it's looking like the long-debated and long-planned natural gas pipeline from the Arctic may never be built, as new plans for liquefied natural gas (LNG) plants and terminals proceed. The opening of year-round shipping through the Bering Strait and via the Northwest Passage will allow LNG to be delivered from around the world more cheaply and to much wider markets.

Just how dumb have we been? There are now nine multi-million-dollar projects in the works to import LNG into Canada from as far away as Malaysia, Indonesia, and Russia. Some of the projects will present very serious environmental and safety concerns.

But considering all of this, and the huge recent petroleum industry profits, surely petroleum firms in Canada are planning to increase their spending on oil and gas exploration and development? Guess what? Estimates show 59 percent of companies surveyed worldwide planned spending increases. But in Canada, 57 percent planned to decrease spending.[26]

If you think we've been incredibly stupid with NAFTA and our oil and natural gas, wait until you read the chapter on water, which follows. But you'd better be prepared to have some blood pressure pills handy.

The September 2007 report by a citizens' panel suggesting that petroleum industry royalties in Alberta be increased produced an entirely predictable, heated, and sometimes hysterical industry reaction. Bill Hunter, chairman of the panel issuing the unprecedented report, which urged higher royalties particularly for oil sands production, said, "Albertans do not receive their fair share from energy development and they have not been receiving their fair share for some time."[27]

The report suggested that there was lots of room for higher royalties, that they wouldn't drive the oil companies out, and that the province of Alberta was missing out on billions of dollars in revenue it should have been getting.

EnCana, the company with the largest corporate profit in Canadian history in 2006, threatened that it would cut $1-billion from its scheduled 2008 spending plans if asked to pay higher royalties. Canadian Natural Resources, ConocoPhillips, Nexen, Talisman, and Imperial Oil, among others, also warned against royalty increases. One broker, in an e-mail to his clients, absurdly called Calgary "Caracas on the Bow River." Meanwhile, the Canadian Association of Petroleum Producers complained that the 104-page royalties report, *Our Fair Share*, ignored the real costs facing the industry, while several analysts were using the title "Albertastan." At the same time, some estimates placed the loss to Albertans because of inadequate royalties in recent years at almost $9-billion.

As the debate continued, it was revealed that the Alberta Energy Department had told the Ralph Klein government in no uncertain terms three years earlier that oil and natural gas royalties could rise significantly without hurting the industry. In a highly critical follow-up report, the Alberta auditor-general said he didn't know why the government chose not to act.

Does anyone think it might possibly have anything to do with the generous petroleum industry election funding going to the Klein government?

For some perspective on this, it's interesting to note that, according to Wood Mackenzie, the Alberta government's net revenues from the oil sands were the 11th lowest as a percentage of industry revenue among a list of 100 governments around the world, and if the province raised royalties as the new report suggested, Alberta would rise on the list, but only to number 44.[28]

The response to the possibility of rising royalties has been so harsh, bordering in some cases on the hysterical, that for the first time in my memory (and I've lived in Alberta for over 70 years) many people in the province are turning against the industry. The excellent *Financial Post* Calgary petroleum correspondent, Claudia Cattaneo, writes that the

> reservoir of goodwill seems to have vanished. A surprising number of Albertans seem to be cheering for the six-member panel calling for a punitive increase in taxes and royalties from Big Oil. Warnings from Big Oil are scoffed at as "posturing." Many, including the panel itself, believe oil companies have no choice but to pay up because they have nowhere else to go since the rest of the world has gotten even meaner.[29]

Eric Reguly points out that "economic nationalism is on the rise. Poor countries that once begged foreign oil companies to develop resources buried in deserts and jungles are now kicking these same companies out. Development revenue and tax laws are being furiously rewritten to favour domestic interests."[30]

Widely respected Calgary journalist Andrew Nikiforuk lauded the royalties report for its "blunt conclusions" and its honesty, calling Alberta's
resource record "a third-rate sham."

When the Alberta government responded to the panel's call for a 26
percent increase in the province's royalties, the response was a remarkable watering down of the proposed increase. The changes were far more
timid than most had hoped for, leaving Alberta still far down the list of
petroleum producing countries in revenue going to government coffers.
Pedro van Meurs, a respected international consultant on royalties,
called the new royalty rates "a disaster for Alberta [which] will not give
Albertans a fair share of the oil sands revenue."[31]

Despite the ludicrous and wildly distorted threats from the industry
("many international investors will say forget Canada and forget
Alberta"), all this has happened as oil prices rose sharply and federal corporate taxes were being chopped. UBS Securities Canada said that the
net value per barrel of oil sands in the ground was now greater than the
value before the modest royalty increases were announced, and projections were that even with the increases and substantially increased production, royalty revenue will fall by over $10-billion over the next 10 years
and Alberta (the "energy superpower") will continue to get more revenue
from gambling than from oil sands royalties.[32]

Lastly, on the subject of energy and the petroleum industry, one of
the funniest things I've read in Canadian newspapers in years was the
October 2007 headline in the *Globe and Mail* for a Doug Saunders dis-
patch from London, England, which read, "Energy CEOs Call for
National Policy." Anyone familiar with the history of the industry in
Canada will know how ludicrous this is given the protracted and bitter
opposition by oil patch executives to any sign of Ottawa getting more
involved in the industry. Now, Patrick Daniel, heavyweight CEO of
Enbridge, of all people, is calling for "a national energy strategy."

It truly boggles the mind. It could be the basis for a great satirical
Canadian play.

WATER

THE COMING CONFRONTATION
WITH THE UNITED STATES

There's one other natural resource to consider before we go on to other matters – water. Prior to the Canada-U.S. free-trade negotiations, incredibly, Canada's chief negotiator, Simon Reisman, was said to be in favour of allowing water exports to the United States, but fortunately many officials in Ottawa and others on the negotiating team were not.

During the lengthy debates in Canada about the FTA in the mid- to late-1980s, Canadians were repeatedly assured that there was nothing to worry about – water was not included in the proposed agreement. There were full-page big-business ads in newspapers across the country that said expressly, "water is not included." Either big business was intentionally lying to Canadians, or, like so many pro-free-trade advocates such as former Conservative International Trade Minister John Crosbie, they hadn't taken the time to study closely the proposed agreement. Amazingly, Crosbie admitted he hadn't even read it.

Water is in fact part of both the FTA and NAFTA because of the definition in the agreements of "goods" as defined in GATT's harmonized commodity coding system, which includes Tariff Item 22.01: "Water: All natural water other than sea water."

Moreover, in Annex 702.1 of NAFTA, any doubt about water being included is removed so that water is clearly "a tradeable good" subject to all the onerous FTA clauses, and to NAFTA's terrible Chapter 11, which, as previously described in my chapter on NAFTA, allows U.S. companies to sue Canadian governments. (In case you think this is just theoretical, the American Sun Belt Water Company filed suit in 2000 against Canada for $200-million because of British Columbia's refusal to allow the export of water to the United States.)

NAFTA Article 309 prevents governments from restricting or prohibiting the export of goods. Moreover, if any favourable treatment is available to Canadian investors, Article 1102 of NAFTA accords American investors the same treatment, which means American companies must be granted national treatment status on a par with Canadian companies. It's important to understand this. NAFTA laws are supernational; that is, they supersede national laws, whether they relate to resources or social policy.

The *Globe and Mail*'s Eric Reguly mentions a little known but crucially important point. During the negotiations,

> water exports were off the table, that is, water was special and would not be treated as a tradeable good like oil. Mysteriously, the exemption did not make it into the final draft.
>
> In short, this appears to mean that once water exports start, they can't be stopped.
>
> Environmental concerns aside – it's an open question whether Canada even has a surplus of water. What's wrong with water exports? Ask the Canadian auto industry or the Alberta oil sands companies or any manufacturer that depends on plentiful supplies of cheap water (most do). Now, imagine that water becomes a tradeable good under NAFTA. All of a sudden Canadian industry would compete for water supplies they had taken for granted.
>
> Water is a vital economic tool. Exporting water is tantamount to exporting jobs. In a world of scarcer and scarcer reserves of fresh water, water is the most valuable economic tool this country possesses.[1]

Any wonder that the former U.S. ambassador to Canada, Paul Cellucci, has suggested that bulk water exports be traded on the open market? As the *Globe*'s Lawrence Martin has suggested, both Trudeau and Pearson "would have politely told him to go jump in the lake."

Next to the United States, on a per-capita basis, Canada is the highest user of water in world, well above the OECD average, and ahead of 28 other OECD nations. We also have the third largest supply of fresh water but only about 6 to 7 percent of the world's renewable supply,[2] much less than the oft-quoted 20 to 25 percent.

In terms of water withdrawal as a percentage of annual availability, Norway is at 1 percent while Canada, along with Finland, Ireland, New Zealand, and Sweden, is at 2 percent. By comparison, the United States is at 19 percent and rapidly headed higher.[3] In short, as almost everyone knows, the United States is running out of water, and global warming isn't going to help.

How vulnerable are our water supplies in relation to potential American NAFTA demands? Economist Erin Weir explains: "Free trade does not force Canada to export bulk quantities of water to the U.S. However, if any provincial government permitted such exports, Americans could use NAFTA to require all other provinces to do likewise. The prospect of water being commodified like other natural resources is of great concern to some Canadians."[4]

That's an understatement. Public opinion polls show, again and again, year after year, that Canadians are overwhelmingly opposed to water exports. Any politician who allows them won't last long in office. Yet as sure as you are reading these words, there will soon be huge pressure from the United States to gain access to our supplies, and some character may see no harm in exporting water. (Someone like Roger Grimes, for example, who, as premier of Newfoundland, planned bulk water exports to the United States until public opposition and financial pressure from Ottawa changed his mind.)

I don't think many Canadians need to have the importance of fresh water explained to them. If you're unsure about the matter, I strongly suggest you read Maude Barlow's excellent book *Blue Covenant: The Global Water Crisis and the Coming Battle for the Right to Water.*[5]

B.C. agrologist Wendy Holm has it right: "There is a global struggle emerging between commodity and community. Commodity is winning. It is time to draw a line in the sand and say 'here is where the rights of commodities end and the rights of communities begin.' Water is that dividing line."[6]

Howard Mann is an international lawyer based in Ottawa who specializes in sustainable development law. He says,

> The key issue of whether trade law can compel states to sell freshwater through diversions, bulk exports, bottling, or other means remains a live one.
>
> This issue became so serious in 1992–93 that Canada demanded an interpretative note from its NAFTA partners.
>
> But, the NAFTA water statement left as much open to question as it possibly could have.
>
> In short, there remains significant uncertainty as to how trade laws will or will not constrain government abilities to prohibit or to restrict exports of freshwater resources. This uncertainty is compounded by elements of international investment law which have led to rulings, in at least three cases in recent years, that the right to export products can be seen as part of the set of protected rights of foreign investors.[7]

I had a long conversation with Howard Mann in April 2007. He is completely clear: Once you start exporting water you have crossed "a tripwire," setting a process in motion that will be very hard to stop. The federal government has failed to do its job of protecting our water because we never properly closed the door to exports. In that conversation, Mann told me,

> We should have clearly and firmly said no! We have the jurisdiction and we should not allow water exports. Sure as hell, a

U.S. government or private firm is going to try to use our existing trade agreements to get access to our water.

Yes, absolutely, Washington will take action itself or encourage private U.S. corporations to do so with the government's backing.

The respected former GATT trade negotiator Mel Clark is equally clear. While under the GATT Canada had complete control of its water, under the FTA and NAFTA, the agreements

transfer control of Canada's water to the U.S. They give Americans the same rights as Canadians to our water and they override the constitutional right of provinces to control water in their territories and accord American corporations the right to sue the federal government and/or the provinces if they fail to respect NAFTA.

NAFTA national treatment is unlimited and there is no exception for water.

Sooner or later, Americans will invoke their NAFTA rights to import water from us. We'll have no right to stop them. The only way to regain control of our water is to terminate the FTA (Article 2106) and NAFTA (Article 2205).[8]

I say we should definitely show the Americans that we *do* respect NAFTA, just as they showed Canadians their respect for NAFTA in the softwood lumber fiasco.

Peter Lougheed, one of the strongest supporters of the FTA and NAFTA, warned in November 2005 that "with climate change and growing needs, Canadians will need all the fresh water we can conserve, particularly in the Western provinces. Water availability is going to rise to the top of the U.S. domestic agenda."[9] Like banker Gordon Nixon on the subject of U.S. direct investment in Canada, Lougheed appears not to understand just how terribly vulnerable Canada has become because of the two poorly negotiated agreements.

As I am writing these words, there is news of more secret closed-door meetings, this time held by the so-called North American Future 2025 Project, which specifically include discussion about water transfers and water diversions as part of the Security and Prosperity Partnership, which will be described further in the conclusion of this book.

"It's no secret that the U.S. is going to need water. It's no secret that Canada is going to have an overabundance of water. . . . There may have to be arrangements." So said Armand Peschard-Sverdrup, the director of the North American Future 2025 Project, which is headed by the Center for Strategic and International Studies in Washington, D.C., in partnership with the increasingly misguided and continentalist Conference Board of Canada in yet another high-level round of behind-closed-doors, big-business-planned talks that are not open to the public or the media.

What has any of this got to do with the will of the Canadian people and the way democracy is supposed to function in Canada?

We must have a new legislated national water policy to ban the export of water. Once there are exports, it will be impossible to turn off the tap, regardless of Canadian needs.

Meanwhile, as the noted Canadian fresh water scientist David Schindler has reported, rivers and lakes on the prairies are drying up, and some river flows are already down between 40 and 80 percent. At the same time, the Munk Centre for International Studies at the University of Toronto is clear: Federal and provincial bans against bulk water exports "probably will not stand up to court challenges" as long as the FTA and NAFTA remain in place.

Canadians had better decide to protect their precious water supplies, or soon it will be too late to do so. If you're apprehensive about any of this now, take a drive and look at what's happened to your favourite glacier.

GROSS DOMESTIC PRODUCT

DEFINITELY NOT TO BE CONFUSED WITH
STANDARD OF LIVING OR QUALITY OF LIFE

I have explained many times elsewhere why frequently used GDP figures can be so very misleading and confusing. I'll come to standard of living and quality of life in the next chapter, but here I'll deal with how our politicians and the media consistently manage to misinterpret the relevance of GDP figures. (See the glossary at the back of this book for GDP and PPP definitions.)

Per-capita GDP rankings are not very valuable if distributions of incomes are badly skewed as they are in Canada. An example I use is this. Suppose there are 11 people in a room, and one earned a million dollars last year, and all the others earned only $10,000. The average income of the 11 would be $100,000. Sounds pretty good, but the reality, of course, would be that one person was wealthy and the other 10 were living in poverty. This same type of distortion occurs with the most commonly used measurement of standard of living, GDP per capita.

I'll review other drawbacks in the use of GDP to measure quality of life in the next chapter, but for now let's review how GDP is reported by our political leaders, by almost all of our economists, and by virtually all of our media.

Depending on how GDP is measured, using purchasing power parities (PPPs), or exchange rate comparisons, or per-capita comparisons, or national GDP totals, Canada comes in anywhere from ninth to 23rd place in a long list of other countries.

In the latest reliable comparative figures at this writing, Canada had the ninth largest economy in the world measured in billions of dollars. However, if you measure GDP per head in terms of purchasing power, we're down in 11th place, below Luxembourg, Norway, the United States, Ireland, Switzerland, Iceland, Denmark, the Netherlands, Austria, and Australia.[1]

Looking at overall GDP growth from 1992 to 2006, Canada was in 13th place among OECD countries. In 2006, 21 OECD countries had a higher GDP growth rate than Canada.

Let's look again, this time by doing a comparison of the 10 years from 1996 to 2005. During these years, Canada's real GDP increased on average by 3.34 percent a year. During the same years, all OECD countries averaged a 2.73 percent increase, the Euro area 2.1 percent, and the average increase for the United States was 3.34 percent, exactly the same as Canada's.[2]

So what can we expect in the future? Some very different forecasts are what we now have. The International Monetary Fund, in its September 2006 World Economic Outlook, said that 135 countries will have grown faster than Canada in 2007. On the other hand, the OECD suggests that Canada will have been in eighth largest GDP position in 2007, comparing the total GDP for each country.

In 2007, our total GDP was forecast to have been less than that of the United States, Japan, Germany, the United Kingdom, France, Italy, and China, all of course with larger populations than Canada.

This said, forecasts for 2020 by the research and advisory firm Economist Intelligence Unit suggest that Canada will be in 12th place in total GDP, behind, in order, China, the United States, India, Japan, Germany, the United Kingdom, France, Brazil, Russia, Italy, and South Korea, with Mexico and Spain essentially tied with Canada.[3] For those who regard this as a terrible trend, note again that these are not per-capita rankings, and all the countries listed above have greater populations than Canada.

Of course, all of these forecasts were made before the disastrous U.S. sub-prime debacle hit the world's economy.

STANDARD OF LIVING
AND THE QUALITY OF LIFE

I t's bad enough when our politicians and journalists confuse GDP per capita with standard of living, but it's even worse when they mix it up with quality of life, as they frequently do. Witness, for example, this remarkable statement by *Globe and Mail* columnist Neil Reynolds: "As befits the country that proclaimed it as a fundamental right, the U.S. is a front-runner in the world's pursuit of happiness. Here you keep score by tracking per capita GDP."[1]

What a truly bizarre statement! The 2007 OECD publication *Society at a Glance* measured the percentage of people who feel an above-average level of satisfaction with their lives. Of the 30 OECD countries, 14, including Canada, show a higher degree of satisfaction than the United States. Of those 14, 12 have a lower GDP per capita than the United States. In *The Economist* magazine's 2007 *Pocket World in Figures,* five of the top 25 cities classified as having the "highest quality of life" are Canadian. Not one U.S. city is among the top 25.

Measuring standard of living or quality of life without considering such things as the quality of social programs, distribution of income and wealth, leisure time, the environment, the amount of violence in society, and the purchasing power of a currency can be, and most frequently is, very misleading.

As noted American economist Paul Krugman has pointed out,

Frances's GDP per person is below that of the United States, but much of that is because the French take more vacations, averaging about seven paid weeks per year, and work shorter hours and prefer to spend much more time with their families or in leisure activities. Yet French schools are good almost everywhere in the country, and there is guaranteed access to quality health care. Yes, it's true that the French have lower per capita incomes than Americans, but who will argue that their quality of life is poorer or that Americans are happier? Or that Americans have superior family values?[2]

For seven years in a row, Canada was ranked first in the United Nations Human Development Index (1994 to 2000 inclusive). While this annual report is far from comprehensive, it does consider such important factors as health, longevity, literacy, and education enrolment, plus GDP per capita at purchasing power parity. Even though in the 2007/2008 edition Canada is ranked fourth, marginally behind Iceland, Norway, and Australia, our score has in fact risen, from 0.870 in 1975 to 0.961 – and we're only 0.07 behind first-place Iceland, which has a score of 0.968.

In the life expectancy index we're ranked fifth. In comparison, the United States is ranked 29th. (The Human Development Index covers 175 UN members plus three other countries.) Every year, Canada ranks well ahead of the United States in these and other quality-of-life comparisons, and ahead of most OECD countries. Despite all of this, if you read a Canadian newspaper, you can read almost every week about how Canada's standard of living has been falling compared to that of the United States. And although they don't often come out and say so in so many words, the implication is clear that our quality of life is also quickly falling behind.

If you believe the reports from many of our business economists and right-wing think tanks, Canada's "standard of living" has fallen from almost 90 percent of the U.S. level to only just over 84 percent in 2005. But in fact it wasn't at all our "standard of living" that was falling, and compared to the United States it sure isn't our quality of life. In fact, what they're using as a comparison is Canada's GDP per capita, which, as already indicated, does not accurately represent either per-capita standard of living or quality of life.

As we have already seen, there is considerable debate about Canada's weak productivity compared to the United States. Some economists blame taxes and regulations, others (including me) blame, among other things, the failure of business to invest in new machinery and equipment and its failure to do adequate R&D.

There is also considerable debate about the impact of productivity on living standards. Heather Scoffield, following an interview with Andrew Sharpe of the Centre for the Study of Living Standards, put it this way in the *Globe and Mail:* "In the world of economics, happiness equals gross domestic product divided by population, but in the real world, happiness is harder to quantify. Countries such as Canada have a vague but broad concept of well-being that goes far beyond GDP per capita."

Vague? I think not. It's called quality of life. And it includes a long list of things which are normally considered necessary parts of "a good life," a list that doesn't include the widespread abject poverty, lack of health care, violence, and incarceration rates that are found in the United States. As Neil Brooks points out, Canada rates well ahead of the United States in terms of the economic security in the population. Moreover, as Brooks says, "Based on the most recent survey data . . . Canadians report they are among the happiest citizens in the industrialized countries."

When's the last time you heard a big-business-funded think tank economist referring to happiness? And for our Americanizers who want even greater "integration" into the United States, a November 2007 Angus Reid poll for *Maclean's* showed that when Americans were asked if they would expect to have a better quality of life if they moved to Canada, a resounding 91 percent said yes.[3]

Commenting on yet another right-wing academic economic analysis that claims Americans are increasingly better off than Canadians,[4] Neil Brooks was once again perceptive.

Aiming for U.S. per capita GDP appears to be totally misguided. On nearly every indicator that we looked at, the
Americans were at or near the bottom. They have the highest
rates of poverty, the greatest degree of insecurity and inequality. Their health outcomes on average are the worst. So are
their education outcomes. They may have more obscenely
rich people than we do, but that doesn't make the typical
family any better off.

At the end of the day we should be concerned about
whether we are enjoying life and leisure.

The OECD elaborates on this use of GDP as an economic or social
indicator:

> GDP excludes a range of non-market activities that influence
> well-being, due frequently to practical concerns with measur-
> ing them, because their value is not easily defined in market
> terms. These include not only illegal activities and home
> activities like housework and do-it-yourself work, but also
> leisure, which is clearly of value to society.
>
> GDP does not take account of externalities, such as pol-
> lution or environmental deterioration, nor of depletion of
> non-renewable resources.
>
> GDP does not distinguish differences in the distribution
> of income. To most people, a huge increase in national
> income that goes exclusively to a tiny handful of very wealthy
> families will not increase general well-being as much as if it
> were more equitably distributed.[5]

As we've already seen, the average American works longer hours than
the average Canadian, a remarkable total of almost two weeks a year more,
according to one survey. But very long hours are judged to be obscene in
most developed countries. When the Innovative Research Group asked

Canadians if they were willing to give up their vacation time, 72 percent said no, and 65 percent said they had no interest in working longer hours. Of course, a country that works shorter hours and has more holiday and leisure time will likely have a lower GDP per capita than a country with longer work hours, but few would suggest that working shorter hours implies a poorer quality of life, and most would think that it meant exactly the opposite.

While Canadians work far fewer hours in a year than Americans, they work more hours than they do in 15 other OECD countries, including Australia, Austria, Finland, the Netherlands, Portugal, Switzerland, Ireland, Sweden, Italy, Belgium, Germany, Denmark, Norway, the United Kingdom, and France. Yet as Andrew Sharpe and colleagues point out,

> Workers in Nordic countries have been able to produce goods and services per capita that slightly exceed the value of the goods and services per capita produced by workers in Anglo-American countries, yet this seriously understates how much better off they are since they are able to produce these goods and services while working over 200 hours less a year. (On average American workers work 274 more hours a year than workers in Nordic countries.)[6]

Do Canadian workers work too few hours? Buzz Hargrove, president of the Canadian Auto Workers, also points out that while Canadian workers work shorter hours than Americans, they work longer hours than Europeans, have fewer vacations, put in more overtime, and generally retire later.

The French take, on average, more than 40 days of vacation a year. The Germans take 28 days, the British 23, while Canadians average 19 days a year. (See Note 10 for legally required paid leave.)

All in all, among OECD countries, Canada is almost exactly in the middle in hours worked, with the 14th highest average hours in a year. In 17 OECD countries, the average worker works fewer hours than the average American worker.[7] This said, from 2000 to 2005, while Canadian workers actually increased their hours by about 1.5 percent, all the following countries, which already had low rates, saw a further drop in hours worked: Sweden, Denmark, Norway, France, Switzerland, Austria, and Germany.

How many hours a week do Canadians work? From 1982 to 2002, the actual hours worked per week (average of all jobs) saw a high of 35.66 in 1989 and a low of 33.93 in 1992. In 2005 it was an average of 33.4 hours.[8] The gap between hours worked in Canada and the United States has been narrowing. By 2005, hours worked per capita in Canada stood at 94.7 percent of the U.S. level, after being at only 88 percent from 1994 to 1999.[9]

Another factor worth considering (and discussed briefly in the chapter on distribution of income) is the fact that while Americans have been working longer hours, their median income had fallen for six years in a row up to 2005. In 2000, median U.S. household income was $49,133. Six years later, it was down to $48,201.

Economist Jim Stanford of the Canadian Auto Workers sums it up well. Even though the average French worker puts in fewer weeks in a year than the average American worker,

> for every hour spent working, the French produce slightly more value-added than the Americans. In fact, most of Europe has higher productivity than the United States.
>
> So, the French work less, but they work more productively.
>
> As a result, they have high incomes, yet spend a lot more time protesting, dining out, and making love (not necessarily in that order).
>
> Now that's what I call efficient.
>
> There's a natural tendency to confuse how well off you are with how much "stuff" you have. Stuff is visible, and has a price tag.
>
> Time, on the other hand, is invisible and apparently free.
>
> Every ranking of countries, according to GDP produced per capita, falls into this trap, by implicitly assuming that

the value of time is zero. (In fact, the older I get, the more I realize time is priceless, not worthless.) . . .

If you work longer, hence boosting your GDP per capita, then by definition you are more prosperous. The most prosperous nation of all would be the one that can find a way to keep people working 24 hours a day, seven days a week. . . .

In other words, most of the much-discussed "prosperity" gap between Canada and the U.S. is due simply to the fact that employed Americans work longer. That's not my idea of "prosperity."[10]

Now, if we could only get the people who own this country to invest more of their record profits in new technology, new machinery and equipment, and in much more R&D, we could enjoy being even more like the Europeans and less like the Americans.

And next time you come across an economist or a politician or a journalist equating GDP per capita with standard of living or quality of life, tell them they can get a copy of *Economics for Dummies* at any good bookstore.

Lastly on this subject, within just a few days of each other, our leading Canadian newspapers ran the following headlines: "Canada-U.S. Prosperity Gap Grows Wider," which began a story featuring the annual report of the University of Toronto–based Institute for Competitiveness and Prosperity;[11] followed four days later by "Canada's Fortune Reversal," showing how our Canadian per-capita income increased at almost twice the U.S. rate between 2002 and 2006;[12] followed three days later by "We're Lagging Behind the U.S. in Productivity."[13]

Confused? You should be, given everything that's been drummed into you about how vitally important productivity is to our standard of living. John Godard of the I.H. Asper School of Business at the University of Manitoba writes:

There is no automatic association between productivity gains and either the wealth or well-being of workers. The question is not just how we can achieve productivity gains, it is also how such gains are to be distributed. Until such time as proponents of the productivity agenda explicitly address this part of the equation, their cries are likely to fall on deaf ears. And so they should.[14]

Bravo to John Godard!

PART NINE

REFORMING OUR DYSFUNCTIONAL
ELECTORAL SYSTEM

SOMETHING THAT SHOULD BE
OUR NUMBER ONE PRIORITY

Wherever I go across Canada, during the question-and-answer session after one of my speeches, someone in the audience invariably asks me what my most important priority would be to make Canada a better country. My answer has been the same for much of the past 30 years. We urgently need election reform.

Remarkably, almost 40 percent of those who didn't vote in recent elections said that the elections didn't matter, or that they felt they had no one to vote for. In a UN list of 179 countries, when voter turnout was calculated as a percentage of all eligible voters, Canada placed way down in an astonishing 93rd place (the United States was 134th).

Poll after poll has shown that the public has become increasingly disillusioned with politicians and politics. Only a small percentage say they trust politicians and feel their votes matter. In one poll, 70 percent said they believed politicians were corrupt. Among the 30 OECD countries, 18 have a greater proportion of respondents than Canada who say they have a high level of trust in parliament.[1] Voters have become more and more cynical and turned off. But why shouldn't they be, given the shenanigans and corruption and broken promises we've seen in Ottawa during the Mulroney, Chrétien, Martin, and Harper years?

(Peter C. Newman says that Canadians have become so cynical about politics that even when Cabinet ministers admit they lied nobody believes them.)[2]

One appalling result of all this is that over 80 percent of adult Canadians have never belonged to a political party, and fewer than 2 percent of adult Canadians make a donation to a political party. In fact, it's worse than that. In the 2006 federal election, only 0.75 percent of eligible voters made a political donation.[3]

How can we begin to change this? I believe we urgently need a mixed-member proportional representation system. Here are just a few examples of the absurdity of the present system (with special thanks to Democracy Watch and Fair Vote Canada).

- In the 2006 federal election, the New Democrats received a million more votes than the Bloc, but the Bloc received 51 seats in the House of Commons while the NDP got only 29.
- The Green Party received over 650,000 votes across Canada yet elected not a single MP. The Liberals in Atlantic Canada received only 475,000 votes and elected 20 MPs.
- The Conservatives in the Prairie provinces got three times as many votes as the Liberals but elected almost 10 times as many MPs.
- In 2006, with proportional representation, the Conservatives would have won 113 seats instead of 124, the Liberals 93 seats instead of 103, the NDP 59 seats, not 29, the Bloc 31 seats, not 51, and the Green Party 12 seats, not zero.

With proportional representation, regional representation would be infinitely fairer, and would much better reflect voter intentions.

Larry Gordon of Fair Vote Canada puts this important topic well:

First-past-the-post has a devastating effect on Canadian unity by exaggerating and exacerbating regional differences. By over-rewarding parties with a strong regional base of support and suppressing . . . minority political views, the resulting electoral map masks the variety of political views held in all regions. Canadians in all regions have a diversity of views, but the electoral map makes Canada look like a hodge-podge of partisan fiefdoms.

Fair Vote Canada (www.fairvotecanada.org) put out a superb report analyzing Canadian federal elections between 1980 and 2000. Among their conclusions:

- In 1984, Brian Mulroney won almost 75 percent of House of Commons seats with 50 percent of the vote.
- In 2000, over 200,000 Alliance voters in Saskatchewan elected 10 Alliance MPs, but in Quebec over 200,000 Alliance voters elected zero MPs.
- Canada ranked a terrible 36th among nations in our percentage of women MPs.
- In 2000, 2.3 million Ontario Liberal voters elected 100 Liberal MPs; 2.2 million other Ontario voters elected only three MPs from the other parties.
- In the six federal elections between 1980 and 2000, over 49 percent of the votes were wasted; that is, they elected no one. In countries using proportional voting systems, only 5 percent of votes in New Zealand were wasted, only 6 percent in Scotland, only 7 percent in Germany.

Some more distorted election results:

- In 1988, Alberta Conservatives won 100 percent of the seats with 52 percent of the vote.
- In 1997, Liberals won all four P.E.I. House of Commons seats with only 45 percent of the vote, and three years later all four with 47 percent of the vote.
- In Quebec in 1980, Liberals won 74 of 75 seats with 68 percent of the vote.
- In the 2006 federal election, hundreds of thousands of Conservative voters in Toronto, Montreal, and Vancouver

should have elected about nine MPs, but instead elected no one.[4]

- In 2004, fewer than 180,000 Conservative voters in Saskatchewan elected 13 Conservative MPs, but over 300,000 Conservative voters in Quebec elected no one.

- In 1993, Kim Campbell's Conservatives won two million votes, but only two seats; that is, only a single seat for a million votes. Meanwhile, the Liberal Party won a Commons seat with an average of 32,000 votes.

- In 2004, more people in Ontario voted Conservative than their combined vote in British Columbia, Alberta, and Saskatchewan. Yet the Conservatives elected only 24 MPs in Ontario compared to a total of 61 in the three Western provinces.

- In 2004, 1.7 million people voted for the Bloc, which won 54 seats in Parliament. In the same election, the NDP received 2.1 million votes and won only 19 seats.

- In 1980, the Liberals received about one-quarter of all votes in Western Canada, but instead of getting 20 seats they elected only two MPs.

- In 1998, the Parti Québecois lost the popular vote in the provincial election, but won the most seats and formed the government.

- In 1999, the Parti Québecois received 42.7 percent voter support, but won 60.5 percent of the seats in Quebec's Assembly. The Liberals won more support than the PQ, with 43.7 percent of the votes, but they took only 38.7 percent of the seats.

- In Ontario, Bob Rae won a majority and 55 percent of the seats in the legislature with only 38 percent of the vote. Dalton McGuinty's Liberal majority of 70 percent of the seats came from only 46 percent voter support.

- In 2001, the B.C. Liberals won 77 of 79 seats with only 58 percent of the vote.

- In the 2006 federal election, the Bloc took more than two-thirds of the seats in Quebec with less than half the vote.
- In 1987, Frank McKenna's Liberals in New Brunswick took 100 percent of the seats in the legislature with 60 percent of the vote.
- In three federal elections, Jean Chrétien never received more than 41 percent support, yet won majorities ranging from 51 percent to 60 percent of the seats in the House of Commons. In 1997, he won a majority in Parliament with only 38.5 percent of the popular vote. In 2000, he won a majority with the support of only one in four eligible voters.

As Fair Vote Canada has pointed out, it took only 30,432 votes on average to elect a Bloc MP in the 2006 federal election, 43,314 votes to elect a Conservative MP, 43,468 to elect a Liberal, and 89,338 votes to elect a New Democrat. And, of course, with 665,940 votes, the Green Party elected no one.

Fair Vote Canada makes another important point:

> Since World War I, Canada has had 15 "majority" governments. In each case, one party held a majority of seats and exercised 100% of the power.
>
> But only 4 of the 15 actually won a majority of the popular vote. Four legitimate majority governments over the past eight decades![5]

There are scores of other similar examples in Canada.

Now let's look at how fair the regional distribution of seats has been in the House of Commons. The answer is not very. Alberta now has 28 MPs, but based on population it should have 33. B.C. has 36, but should have 43. Ontario has 106, but should have 120. (While Stephen Harper has proposed partially rectifying this situation, his formula still leaves Ontario badly short-changed by 11 fewer seats than its population warrants, and the 22 proposed additional Commons seats in B.C., Alberta, and Ontario would not come into effect until 2014.)

Toronto has a population larger than all four Atlantic provinces put together, but Toronto has only 23 MPs and the four provinces have 32.

Does the unfairness of the system turn people off? Judge for yourself. One survey said that of the 40 percent of Canadians under 30 years of age, only 5 percent belonged to any political party. In 2003, polls showed that only 3 percent of 22- to 29-year-olds worked as a volunteer for a political party.

Now let's return to the question of voter turnout.

How do Canadians feel about our democracy and the way it functions? In April 2006, in a Leger Marketing poll, only 36 percent of respondents said that Canada is governed by the will of the people. If this is how people feel, is it any wonder that in recent years voter turnout has been declining? In 1958, 1962, and 1963, it was over 79 percent. In 1984 and 1988, it was down to 75.3 percent of eligible voters. In 1993, it was down again, to 69.6 percent, and down once more in 1997, to 67 percent. And in 2000, it dropped to a low of 64.1 percent, though it was back up slightly to 64.7 percent in 2006.

How well does our democracy function? In 2000, only 25 percent of eligible voters voted for the winning party, while 39 percent did not bother to vote. In 2006, the Harper government was elected with 36.3 percent of the eligible vote.

Of course, aside from our highly unrepresentative voting system, there are other important reasons for low voter turnout, including the corruption, disillusionment, and disappointment of the Mulroney, Chrétien, and Martin years, already mentioned, plus the seemingly perpetual list of broken political promises and the perception (correct for so much of the past) that a wealthy plutocracy's big money has been calling the shots in politics in this country.

How about young people, "the future of our country"? Between 70 and 75 percent of 18- to 24-year-old registered voters don't bother to vote. Why don't young people vote? In 2000, 38 percent of eligible young people said they didn't vote because of lack of interest, and 41 percent between the ages of 25 and 34 gave the same answer. Asked to name the most important issue in an upcoming federal election, almost 30 percent said either that there wasn't one or that they didn't know of one. Of the non-voters, 37 percent said either that the elections didn't matter or they didn't like the choices.

All in all, a dismal picture.

In one poll of adult voters, when asked if big business was in charge of our politics because of globalization, 74 percent agreed. When asked if political leaders told the truth and kept their promises, 73 percent said no – the same percentage that said government didn't care about average people. In a March 2005 Leger Marketing poll, when Canadians were asked who, out of a long list, they trusted, they put politicians dead last, at only 14 percent.

How does voter turnout in Canada compare with other countries? I've already mentioned one comparison that put us in 93rd place. But in May 2004, Fair Vote Canada reported that for the decade of the 1990s, Canada ranked all the way down in 109th place in voter turnout.

109th place!

In 2005, 84.5 percent of eligible voters voted in the Danish election. In 2007, 85 percent voted in the French presidential election. Seven other Western democracies have voter turnout rates in the 80 to 90 percent range. All of the following countries have had voter turnouts of 80 percent or better: South Africa, Costa Rica, Argentina, Chile, Guyana, Uruguay, Israel, New Zealand, Papua New Guinea, Belgium, Bulgaria, Italy, Malta, Slovenia, Spain, and Sweden.

Now to some good news. Some *very* good news. An amazing revolution has taken place concerning money in politics in Canada, a revolution that is likely the most important development in political reform to take place in this country in generations.

The profound impacts of Jean Chrétien's unexpected Bill C-24, which severely limits corporate and trade union political donations beginning January 1, 2004, remain to be fully recognized in much of the media, and hence by most Canadians. But it's certainly now fully understood by federal politicians, most notably the Liberals, who have long counted on the big banks, the big law and accounting firms, the big oil and other resource companies, and the wealthy in general to finance their party. What a shock for them that the Conservatives have now been pulling two, three, or four times as much money from roughly 10 times as many donors. Moreover, as a result of the new inflation-adjusted public subsidies based on the number of votes received in the 2006 election ($1.8725 per year per vote at this writing), the Conservatives received $9.4-million in public funding, $1.553-million more than the Liberals, while the NDP received $4.54-million.[6]

The new financing rules have had a huge impact on party financing. The Liberals have suddenly found that the "automatic" big-dollar corporate fundraising via "bag men" that they relied on so successfully for so many years is no longer possible. In 2005, 15,187 individuals donated to the Liberal Party, but almost four times as many people, 59,159, donated to the Conservatives. In 2006, the Conservatives recorded donations from the public of $18.6-million, the Liberals only $9.8-million.[7] In the first nine months of 2007, the Liberal Party raised only $2.6-million compared to $12.1-million raised by the Conservatives.

(Chrétien's completely surprising 2004 political financing legislation was one of only three really important things that he did while prime minister, the others being keeping Canada out of the Iraq war and the clarity legislation regarding future Quebec referenda.)

This said, all is not perfect. Unlimited secret personal donations are still allowed for leadership and constituency candidates as long as it is prior to acclamation, and while political spending is limited during a federal election there are otherwise no limits. (As well, there is a major loophole in the existing legislation which allows an individual to make contributions of $199.99 to every federal riding – a total of more than $60,000 – since riding associations are required to notify Elections Canada only when they receive donations of $200 or more. This loophole should be, and easily can be, closed.)

In March of 2005, I wrote to my e-mail list:

Canada has moved away from the terrible U.S. system where big corporations and single-issue lobbies ignore funding legislation by channelling enormous amounts of money for politicians through the PACs and the successful re-election rate of incumbents is laughable. If we could combine our own funding legislation with a mixed-member proportional representation system we'll have gone a long way towards making Canada the kind of true democracy most of us want to see.

So, ignore those who suggest we go back to dark-age political financing rules. We've had enough of defence contractors, oil and forestry companies, law firms, accountancy firms, and the wealthy financing politicians with the inevitable results that our tax policies and our social policies fail to reflect the true wishes of our citizens as clearly spelled out, year after year, in poll after poll.

It's time progressive policies not fat bank balances determined political leadership. Do we really want another Paul Martin or for that matter another Brian Mulroney and their big-money friends running the country?

The example of Paul Martin's almost $12-million war chest in the campaign for the Liberal leadership was a perfect example of money from big business and big professional firms warping the democratic process. Martin's millions, mostly from big corporations and their CEOs, made it virtually impossible for anyone else to mount an effective, competitive campaign. For Martin, $5,000-per-person cocktail parties were a norm. In March 2006 in Toronto, 2,000 people attended a Liberal fundraiser. Tables for 10 sold for $8,000. Most tables were purchased by large corporations, and a few by trade unions. The evening netted the Liberal Party $1.4-million.

So, in the past, before the recent reforms, what did it take to make a serious bid for the Liberal leadership? Probably a minimum of $3-million to $5-million, and often much more was spent and not reported. Belinda Stronach is reported to have spent $4-million on her campaign for the leadership of the Conservative Party. The most recent Liberal leadership convention, in December 2006, had a supposed $3.4-million per candidate spending limit.

To the total surprise of almost everyone, in 2006 Stephen Harper not only continued the Chrétien ban on all donations from corporations and trade unions, but he also lowered the maximum donation from individuals from $5,000 to only $1,000. As of January 1, 2007, individual Canadians could donate up to a total of an inflation-adjusted $1,100 to each registered political party, and a maximum of an additional $1,100 per year to a registered political association or to contestants for nominations or constituency candidates, whether representing a party or running independently, plus a maximum of $1,100 to a party leadership candidate. This move by Stephen Harper astonished many observers. Equally surprising is the fact that Harper hasn't abolished the tough limits on third-party election advertising and election donations, something he vigorously opposed before becoming prime minister. As well as corporations and trade unions, groups and associations can no longer make political donations, and cash donations exceeding $20 are no longer allowed.

The *Toronto Star*'s Carol Goar concludes:

> No future prime minister will be beholden to corporate interests, as every national leader from Sir John A. Macdonald to Paul Martin has been to some degree. Big Business will never be able to mount another massive blitz, as it did in the 1988 election, to promote free trade with the United States. The era of $5,000-a-plate political fundraisers (at least at the federal level) is over.
>
> Because this story has no single hero and no sudden breakthrough, it hasn't made headlines. But from a citizen's point of view, it ranks as one of the most positive – and surprising – developments of 21st century politics.
>
> Who would have believed, at the dawn of the millennium, that Chrétien, who raked in millions at fundraising dinners, would pull the plug on corporate donations?

Who would have believed the Liberal party, whose president Stephen LeDrew denounced Chrétien's plan as "dumb as a bag of hammers," would enact the ban?

Who would have believed Harper, champion of free markets and opponent of government regulations, would retain and strengthen Chrétien's reforms?

Who would have believed Broadbent, supported by the autoworkers for 14 years as NDP leader and 21 years as MP for Oshawa, would propose an end to political donations by unions?

Most of all, who would have believed that a political system tainted by the sponsorship scandal, infected by public cynicism and dependent on corporate largesse would be getting cleaner and more transparent by 2007?

There is still work to be done.

An area that needs tidying up is the definition of a political contribution and a campaign expense. As long as big-ticket items such as polling and convention fees lie outside the rules, there will be room for slippage.

On balance, though, Canada has come a long way in a remarkably short time.[8]

By the way, the United States is going in exactly the opposite direction. In a country that already had the most undemocratic big-money-dependent system of election financing, the U.S. Supreme Court in June 2007 further relaxed the ludicrously ineffective regulations relating to corporate and union political spending. The cost of the combined 2008 U.S. presidential and congressional elections is expected to break the $1 billion mark.

After the 2006 federal election, Fair Vote Canada said, "Once again Canada's antiquated first-past-the-post system wasted millions of votes, distorted results, severely punished large blocks of voters, exaggerated regional differences, and created an unrepresentative Parliament."

While at the federal level the reforms in political financing have been profound, the prospects for reform of the voting system, despite promises to the contrary primarily from the Martin Liberal government, are at this writing about as remote as the chances of the Toronto Maple Leafs winning the Stanley Cup.

Journalist John Ibbitson sums it up: "To placate the NDP in this minority Parliament, the Conservatives promised in their Throne Speech to consider the question of electoral reform."[9] We now know how they planned to proceed. Those plans are hilarious. The Conservatives hired a Winnipeg-based conservative think tank to conduct focus groups. This same think tank has published articles arguing strongly against proportional representation and in favour of the status quo. Ibbitson says, and I agree, "Mr. Harper has not the slightest interest in considering the question of electoral reform. This charade is an act of political subterfuge calculated to disguise inaction."

As Tom Kent, a distinguished professor at Queen's University School of Policy Studies, has pointed out, the last Chrétien majority "was conferred by only one in four of the people who in 2000 were entitled to vote. In 2004, barely one in five of enfranchised Canadians cast Liberal votes, but that was enough for the party to enjoy the perquisites of office."[10]

Now let's turn to proportional representation (PR).

I've been saying for some 30 years that if we could only get a reformed political financial system that prevented corporations, trade unions, and the wealthy from dominating the financing of federal politics, and if we could combine that with a proportional representation electoral system, we'll have gone a very long way down the road to making Canada a much better, fairer, more democratic country.

Well, we're halfway there, and many of us who worked hard for the reform of election financing really didn't believe it would ever happen. And certainly not with legislation introduced by a Jean Chrétien government. This said, once, to our surprise, we learned of the government's intentions, we went to work as hard as we could to convince many skeptical members of Parliament that the changes being considered could be improved, could make Canada a much more democratic country, and would have the strong support of a large majority of Canadians.

I spent weeks working on my presentation to the House of Commons committee. When the full House of Commons and the Senate approved the new legislation I was overjoyed. And when Stephen Harper brought in an even tougher package of reforms, I was astonished and delighted.

The next half, proportional representation, will likely be more difficult. While Jean Chrétien surprised us all with Bill C-24, it should be remembered that in 1984, when running for the leadership of the Liberal Party, he told reporters he would introduce proportional representation "right after the next election" if he became prime minister. Fair Vote Canada tells what happened: "In 1993, Jean Chrétien wins the election and begins his ten-year reign as prime minister. In three elections, he never wins more than 42 percent of the popular vote, but still forms 'majority' governments thanks to the current voting system. He never gets around to introducing proportional representation."

Let's look at what happens in other countries.

Currently, proportional representation exists in over 80 countries, including Argentina, Austria, Belgium, Bolivia, Brazil, Chile, Colombia, the Czech Republic, Denmark, Finland, Germany, Greece, Hungary, Iceland, Indonesia, Ireland, Israel, Italy, Luxembourg, Mexico, the Netherlands, New Zealand, Nicaragua, Norway, Panama, Paraguay, Peru, Poland, Portugal, Serbia, Slovakia, Slovenia, South Africa, Spain, Sweden, Switzerland, Turkey, Uruguay, and Venezuela.

Pure proportional representation would have the number of those elected as an identical reflection of the percentage of the popular vote received. For example, if party X received 36 percent of the vote in a national or provincial or state election, it would get 36 percent of the seats in the parliamentary chamber.

But most countries have a mixed-member proportional representation system, where the voter casts two votes, one for the constituency candidate they favour, and the other for one of the party lists submitted by the political parties. This system was proposed almost 30 years ago by the Pepin-Robarts commission on national unity, and more recently, in 2004, by the Law Commission of Canada after a careful two years of research, consultation, and discussion. Several provinces, including New Brunswick and Quebec, have been encouraged to adopt mixed-member proportional (MMP) systems for their legislatures.

The Law Commission, in a detailed proposal, suggested that two-thirds of all MPs be elected directly by their constituencies and the balance by proportional representation, using party lists to select the winners, based on the total number of votes received.[11]

There are variations to be considered in a mixed-member proportional system. Germany, New Zealand, Scotland, and Wales have 60 percent of the seats reserved for constituency representatives, and 40 percent are selected from party lists. Most countries with a proportional representation system require a minimum of 5 percent of the vote to elect someone to their parliament.

Studies show that countries with proportional representation have higher voter turnouts. One study suggested that with proportional representation 1.5 million more Canadians would be casting their vote. Fair Vote Canada reported in May 2004 that Canada had far more people not voting than voting for the winning party.

Fair Vote Canada says,

> Based on the large number of countries using proportional or fair voting systems over extended periods of time, international experience demonstrates the following benefits over winner-take-all systems:
> - Wasted votes and distorted election results are reduced.
> - Phoney majority governments are rare.
> - Voter turnout tends to be higher.
> - Parliaments are more representative of the range of political views and the composition of the electorate (gender, ethnicity, regions).
> - These countries maintain strong economic performance.
> - Citizens tend to be more satisfied with the way democracy works.[12]

Alas, if political reform in Canada is a serious problem that must be addressed, the appalling lack of political knowledge in our country is equally disturbing. I briefly touched on this in the chapter on education. For your further consideration, as if the frequently demonstrated ignorance of our history wasn't enough, a 2005 study showed that only 40 percent of voters could define the differences between left and right in politics. Good grief!

In one public opinion poll, almost 60 percent of Canadians thought that aboriginals in this country were either better off than or had about the same standard of living as most Canadians. In a December 2005 poll reported in the *Globe and Mail,* 59 percent were unable to identify our only female prime minister, 43 percent couldn't identify the NDP leader, and 44 percent didn't know which prime minister was responsible for free trade.

In their previously mentioned study for the Canadian Centre for Policy Alternatives, Neil Brooks and Thaddeus Hwong reported that in Canada 11 percent of the population has frequent political discussions with friends, compared to over 20 percent for Greece, Austria, Germany, Denmark, Norway, and Sweden, and an OECD average of 15.5 percent. In a list of 19 countries, only two have a lower percentage than Canada.

A few words about the Senate. Here, I'm entirely with the abolitionists. John Baglow, who has been active on the national council of Fair Vote Canada, writes: "Do we really need an Upper House? New Zealand does quite nicely with only one. The Senate is seen by many as a creaky, elitist institution used to reward the political friends of the party in power."[13]

With a new PR system, we're going to need some more members of Parliament to do the system justice. Let's get rid of the Senate and expand the elected House of Commons. About the dumbest idea I've heard from Stephen Harper is the guaranteed gridlock-inducing election of Senators, a concept fraught with terrible potential problems of conflict, confusion, and rigor mortis.

Yes, yes, I know all about the constitutional problems entailed in abolishing the Senate. So let's have a period of public discussion and debate, and then a national referendum on the subject. I suspect a strong majority of Canadians would approve of abolition. Can one in 10 Canadians name a significant accomplishment of the Senate? And as it is, the representation in the Senate is wonky. For example, British Columbia and Alberta have almost a quarter of Canada's population but only 12 seats in the 105-seat Senate. Atlantic Canada has 30 seats with less than a third of the population of B.C. and Alberta.

Who needs it?

I hope that Tom Kent is right that "public opinion will before long compel electoral reform." Maybe, but when? That the NDP has not pushed proportional representation as a top priority amazes me. It would so benefit the party, and there's little doubt that with an adequate information campaign, a majority of Canadians would be supportive. But without strong government support, which is unlikely, proportional representation or other election reform is difficult. Witness the narrow technical setback in British Columbia in their 2005 referendum on a single transferable vote (STV) system despite majority approval by voters.[14] Note that in B.C. every household received a brochure outlining the STV system. In 2007 in Ontario, no such thing happened.

The electoral reform proposal put before the people of British Columbia in 2005 received majority support in 77 of B.C.'s 79 constituencies but failed because only just under 60 percent (57.7 percent) of voters approved of the proposed change. Some of those who voted against the STV proposal were swayed by the calculation that a seat in the legislature could be won with less than 17 percent of the vote, while many others thought STV was too complex.

Two countries, Ireland and Malta, employ the STV electoral system, plus Scotland for local elections. The Australian state of Victoria held its first STV election in 2007, and Minneapolis will use STV for its municipal elections. Vancouver political columnist Bill Tieleman is strongly opposed to STV.

> STV is complicated, confusing, prone to errors and delay, it reduces local accountability, increases the geographic size of ridings, allows elected members to avoid direct accountability

for their decisions, increases political party control and allows special interests to dominate party nominations.

It also has not been proven to do many of the things its proponents claim – it does not increase the ability of third parties and independents to get elected, and it is not truly proportional in guaranteeing that each party will get the number of seats in the Legislature equivalent to the percentage of votes they received.[15]

Without question, the Canadian people should be allowed to vote in a national referendum on a clear electoral reform question, but only after a good period to allow for a sufficient informed debate.

Electoral reform has also been rejected by 64 percent of voters in P.E.I., and it's not clear that New Brunswick will proceed with their planned 2008 referendum. According to the *Globe and Mail*, "Basic though it is, electoral reform is one of those Important Topics That Cause Eyes to Glaze Over."[16]

Perhaps, but polls repeatedly show that Canadians are in favour of proportional representation. Clearly, what is needed is strong political leadership on the issue. However, anyone waiting for Stephen Harper or Stéphane Dion to provide such leadership is going to be bitterly disappointed, unless citizens across the country apply a great deal of well-organized pressure.

Briefly, let's look at the October 2007 provincial election in Ontario, where Dalton McGuinty and the Liberals won two-thirds of the seats in the election with only 42 percent of the vote, "a resounding victory" according to some in the press. Meanwhile, only 37 percent of voters supported the mixed-member proportional representation (MMP) system recommended by the Ontario Citizens' Assembly, which had met regularly for eight months before bringing forward the suggested reforms. More than 1.5 million Ontarians voted for the MMP in what the press called "a resounding defeat."

What happened? It's not difficult to explain. Many, if not most, didn't understand what they were being asked to vote about. The public education program was totally inadequate. The *Globe and Mail*'s Roy MacGregor put it this way just before the election: "Half of Ontario voters don't know about the proposals. Just ask a few people in a shopping mall. Shrugs. Embarrassment."[17] Larry Gordon of Fair Vote Canada called the public education campaign "pathetically inadequate" and pointed out that more people voted for MMP than for three of the four major parties. He also pointed out that if there was a vote among those aged 18 to 24, the MMP referendum would have easily exceeded the threshold required for approval.

The results from all of this were déjà vu all over again. The 58 percent of voters who voted against the McGuinty government received only a third of the legislative seats. The voter turnout of 52.7 percent was the lowest in Ontario's history.

It's worth mentioning that critics of proportional representation constantly complain that it would result in minority governments. Yet in all the federal elections going all the way back to 1921, only four times was a majority government elected with a majority of the vote.

Lastly, on the topic of electoral reform, it should be remembered that even though Ontario's Dalton McGuinty promised back in 2004 that "decisions must be perceived to be beyond the influence of political contributors," this is the province where individual donations of up to $16,800 were allowed in the 2007 election.

WOMEN IN CANADA

GOOD NEWS AND BAD

How do women in Canada compare with women in other countries when it comes to disparities between the sexes in human development? There are two conflicting lists.

In one, the UN's Gender-related Development Index, Canada comes fourth in a list of 140 countries, behind only Iceland, Norway, and Australia.[1] Of the other G7 countries, France is 7th, the United Kingdom 10th, Japan 13th, the United States 16th, Italy 17th, and Germany 20th. But in a second list, the annual World Economic Forum ranking of gender equality, Canada is down in 18th place, far behind all the Scandinavian countries but well ahead of the United States, which is in 31st place.[2]

When it comes to politics, as we shall see, Canada's performance can best be described as awful.

First, though, let's look at women in Canadian universities. In the 2004/2005 academic year, there were 585,200 women students registered (58 percent of total registrations), compared to only 429,000 men. In master's programs, women accounted for 53 percent of enrolment, but for only 47 percent of doctorate registrations.[3] Women now make up the majority of students in medicine, law, in the arts, in the sciences, and in physical and life sciences technologies, but only about one quarter of those enrolled in architecture and engineering.

While Canada does quite well in the ratio of female to male students when all forms of post-secondary education are included, 10 countries have a higher ratio: Norway, Iceland, Sweden, the United States, Denmark, the United Kingdom, New Zealand, Slovenia, Barbados, and, somewhat surprisingly, Kuwait.

In eight OECD countries – Canada, Finland, Iceland, Ireland, Poland, Portugal, Sweden, and the United States – the educational attainment of females aged 25 to 64, measured by the number of years of schooling, is higher than that of men. In Canada, in 2006, 29.9 percent of females had a post-secondary certificate or diploma, exactly the same percentage as males.

Females are now more likely to complete upper secondary education than males in almost every OECD country. Today, only in Turkey are graduation rates for females below those for males.[4]

In 1990/1991, only 19.6 percent of full-time university faculty in Canada were women, and only 7.6 percent were full professors. By 2002/2003, these numbers improved, but only to 29.9 percent and 17.2 percent, and while they are still far too low, they continue to improve. Since the mid-1980s, however, full-time female professors in Canada have earned some $5,000 less than their male counterparts, an improvement over the 1960s, when the gap was $10,000 to $15,000.[5]

According to Statistics Canada,

> A woman with a university degree in 1977 earned $1.88 for each dollar earned by a woman with a high school diploma. The corresponding ratio for men was $1.63.
>
> By 2003, women with a university degree earned $2.73 for every dollar earned by those with a high school diploma. The corresponding ratio for men was $2.13.
>
> Between 1993 and 2003 the university premium for women was 22% higher than for men.
>
> In 1977, there were four people attending universities from families in the top fifth of the income distribution for every person attending from the bottom fifth. By 2003, this ratio had fallen to only 1:6 for women and to 2:7 for men.[6]

Today, the number of females graduating from law schools in Canada is some 50 percent higher than the number of males, and some 57 percent of students in the first year at medical school at the University of British Columbia are female, as are 60 percent at the University of Toronto and some 75 percent at the University of Montreal. In fact, in 14 of 17 Canadian universities, there are more women than men in medicine. This is quite a change from 1970, when only about 18 percent were women.

In 2004, 34 percent of doctors in Canada were women, somewhat below the OECD average of 38 percent. Of the 30 OECD countries, only Switzerland, Austria, the United States, Luxembourg, Iceland, and Japan had lower percentages of female physicians. The Slovak Republic was the highest at 57 percent, followed by Finland, Poland, the Czech Republic, and Hungary, all above 50 percent. The Nordic average was 42.5 percent, the United States was at only 23.4 percent females.

In 1976, only 42 percent of Canadian women aged 15 and over were part of the paid workforce. By 2006, the percentage had increased to over 62 percent. This percentage of females who were in the Canadian labour force in 2006 compares with the OECD average of 62.8 percent, but it is well below the rate for the Nordic countries: Denmark at 75 percent, Norway 75.4 percent, and Sweden 75.6 percent. Canada's rate of female labour force participation is lower than the G7 average of 66.6 percent and slightly below the EU15 average of 63 percent.

In comparison, the proportion of men who were in the labour force in 2006 was 72.5 percent. Statistics Canada reports that in 2006 almost two-thirds of all the employment gains were among adult women, and the proportion of women aged 25 and over who were working hit a record high in December 2006, bringing their unemployment rate to a 30-year low of 6.1 percent compared to 6.5 percent for men.[7]

In 1987, 30 percent of those employed in managerial positions were women. By 2004, this was up to 37 percent.

In 2003, women working on a full-time, full-year basis had average earnings of $36,500, some 71 percent of what their male counterparts earned. The same year, 31 percent of unattached women aged 16 and over lived on a low income, and 38 percent of all families headed by lone-parent mothers had incomes below the after-tax low income line. In a list of 21 OECD countries, in 2004 Canada had one of the largest gaps in women's full-time earnings compared to men's earnings. Only Switzerland, Germany, Japan, and Korea had larger gaps. Canada, at about 22 percent, had only a slightly larger earnings gap than the United States but was some 4 percent higher than the OECD average.[8]

In 2003, 43 percent of all children in a low-income family were living with a lone female parent, whereas these families accounted for only 15 percent of all children aged 17 and under.[9]

Also bad news is that women in Canada have only a very poor 15 percent of the senior executive positions in business, and, in 2006, only 13.5 percent of all directors on the boards of the 100 largest corporations in Canada were women. Almost forty percent of board directors in Norway are women. Meanwhile, despite the fact that increasingly women have become university presidents and deans of faculties across the country, and while in 2004/2005 about one-third of all university faculty were women, they made up only 19 percent of the full professors. Female full professors earned some $6,000 a year less than their male counterparts. Recent comments from Janice Drakich, director of faculty recruitment at the University of Windsor, described the situation as "systemic discrimination alive and well in the academy."[10]

In terms of self-employment rates, 20 OECD countries have a higher female rate than Canada's 8.1 percent self-employed out of total female employment. This compares with the OECD average of 13.9 percent and the EU15 average of 10.8 percent. The U.S. rate is only 5.9 percent, the worst in the OECD.[11] In 2006, there were 1.62 million men in Canada who were self-employed, compared to 877,000 self-employed women.

In 1982, some one in seven of the top 5 percent of income earners were women. By 2004, it was one in four.[12]

Now let's turn to Canada's poor record on women in politics. As Rosemary Speirs, former national affairs columnist for the *Toronto Star*, writes, "Why is it that our city councils, provincial legislatures and even our House of Commons are dominated by a pinstriped sea of men – an almost 80 percent male political majority? Or, in the four Atlantic provinces, an 85 percent male majority?"[13]

Despite all the talk and all the years of promises, women are not making progress in the numbers of seats they hold in the House of Commons. In the federal elections of 2000, 2004, and 2006 an average of just under 21 percent of those elected were women. In 1993, the number of female candidates was actually higher than in the 2004 and 2006 elections. In the 2006 election, only a very poor 12 percent of Conservative candidates were women and only 26 percent of the Liberal candidates were women. The NDP had 35 percent female candidates. At this writing, in preparation for the next federal election, 42 percent of nominated NDP candidates are women and 35 percent of Liberal candidates, but only a tiny 14 percent of Conservative candidates are women.

In the 2006 federal election, in the 32 ridings in Atlantic Canada, the Liberals nominated only two women, the Conservatives only one, and only one woman was elected, Alexa McDonough of the NDP.

Fair Vote Canada compares the three remaining major democracies still using single member plurality, or first-past-the-post, electoral systems in 2006 in terms of the percentage of women in the lower or single houses:

Canada	20.7%
United Kingdom	17.9%
United States	14.3%

In sharp contrast, here are the percentages of women in major Western democracies using various forms of proportional representation:

Sweden	45.3%
Finland	42.0%
Norway	37.9%
Denmark	36.9%
Netherlands	36.7%
Spain	36.0%
Belgium	34.7%

Austria	33.9%
Iceland	33.3%
Germany	31.8%

After the 2006 federal election, Canada ranked way down in 36th place among nations in its percentage of women MPs, and down in 17th place among the 30 OECD countries. Among the many countries with a higher percentage of women MPs are Rwanda, Mozambique, Portugal, Argentina, Lithuania, Costa Rica, Cuba, Mexico, Belarus, Tunisia, Honduras, South Africa, Pakistan, and Uganda.[14] All of these countries have parliaments that are between 33 and 49 percent female.

Looking at the percentage of women in government at the ministerial or deputy-ministerial level, Canada was down in 31st place at only 23.1 percent as of January 1, 2005. Finland led the way at 60 percent, followed by Spain at 50 percent, Germany 46.2 percent, Norway 44.4 percent, and South Africa 41.4 percent. The United States is far down the list at only 14.3 percent. Other poor performers, also at 14.3 percent, were Luxembourg and Switzerland. The Russian Federation was at an appalling zero percent, as were Singapore, Cyprus, Slovakia, Kuwait, and Uruguay.[15]

I can't leave the subject of the status of women in society without mentioning that 28 years ago the United Nations adopted an important Convention on the Elimination of All Forms of Discrimination Against Women. By 2005, 170 nations had endorsed the treaty. The United States is the single developed nation that incomprehensibly opposes it.

There is very little doubt that Canada's failure to provide a public national child-care system remains a barrier that has held many women back from active political participation during their child-rearing years. Meanwhile, most OECD countries have developed successful policies to increase women's labour market participation and to reconcile work and family responsibilities on a basis more equitable for women.

All Canadians should be ashamed at our failure to do so, and should be ashamed at our embarrassingly low level of female political electoral participation. It's hard to believe we're still way down in an appalling 39th place. How truly disgraceful!

Finally, in 2006, the life expectancy at birth for women in Canada was 83 years while for men it was only 77.4 years. Three countries had longer life expectancies for women: Japan, with 85.6 years, Spain 83.8 years, and Switzerland 83.7 years. Australia, Belgium, France, Iceland, Italy, and Sweden all had female life expectancy rates similar to Canada's. U.S. women were down in 22nd place in life expectancy among OECD countries at 80.1 years.

At this writing, around the world there are 15 elected women as heads of government, including, since 2005, Chile, Finland, Argentina, Jamaica, and the Republic of Korea, and in 2007 a woman was elected in India to the largely ceremonial position of president.

Overall, women in Canada have clearly come a long way in recent years in their level of education and participation in the workforce. Equally true, compared to many other countries, Canadian women still have a long way to go to reach the income and societal levels they deserve.

INTEGRATION BY SECRECY AND STEALTH

"If it isn't a conspiracy they're doing
their best to make it appear like one."

In February 2003, the vice-president of the Canadian Chamber of Commerce, in Canadian Senate hearings, said that de facto integration of Canada with the United States was here already, "whether many Canadians realize it or want to accept it," and that we don't need "duplicate systems of approval" anyway.

Got that? No need for duplicate systems of approval. I suppose, then, there's no need for the House of Commons or provincial legislatures either. Let's just rubber-stamp American policies, standards, and values.

Back in 2002, the continentalist *Financial Post* columnist Diane Francis said that "Canada is more integrated with the United States economically than any two European countries are,"[1] and the noted Canadian economist Richard Harris and the Carnegie Endowment both said the same thing.

David O'Brien, chairman of the board of the Royal Bank of Canada, has said that Canada would have to adopt U.S. immigration policies: "We're going to lose increasingly our sovereignty, but it's necessarily so."[2] Patrick Daniels, president of Enbridge, hilariously complains that Canada pushes its sovereignty a little too far.[3]

In June 2006, the North American Competitiveness Council (NACC), consisting of business leaders from each of the three NAFTA countries, was formed "to advise governments." Toronto lawyer Paul Bigioni called it "an anti-democratic institution."

On the surface, the NACC appears to be an initiative of government. It is not. It was entirely conceived by the private sector. In 2003, the Canadian Council of Chief Executives launched a sales pitch designed to convince governments to pursue such business-friendly initiatives as "re-inventing" borders, regulatory convergence and energy integration. It is no coincidence that the NACC currently pursues the same objectives.

In fact, all the Canadians on the NACC are members of the powerful Canadian Council of Chief Executives (CCCE) lobby group, and the CCCE serves as the Canadian secretariat. The big-business NACC is the only non-governmental organization making recommendations to the secret tripartite Security and Prosperity Partnership (SPP).

In the preface to this book, I alluded to the top-secret meetings – behind closed doors – that have already been held, with more of the same planned for the future. These meetings are designed to further integrate Canada into the United States, have us adopt even more American standards, values, and policies, and give Americans even more guaranteed access to our resources and the unimpeded ability to buy up the ownership and control of even more of our country.

Perhaps you don't know about the three days of highly secret meetings that took place at the Banff Springs Hotel in mid-September 2006, meetings between top-level American, Canadian, and Mexican government officials and many senior corporate heads. In fact, you probably *don't* know. But then again, why would you know? Despite the fact that the list of very high level attendees was leaked to me and I sent it to the media, along with the agenda, there wasn't a word about the meetings in our two national newspapers, the *Globe and Mail* and the *National Post*. There was nothing on CBC television, on CTV, or on Global.[4]

The documents that I obtained and sent out had been marked "Internal Document. Not for Public Release." The three heavyweight co-chairs of the secret meetings were former Alberta Conservative premier and strong pro-FTA advocate Peter Lougheed, former U.S. Secretary of

State George Shultz, and former Mexican Secretary of the Treasury Pedro Aspe.

Among the many well-known Canadians scheduled as "participants" were Stephen Harper's Conservative cabinet ministers Stockwell Day (who at first denied attending) and Gordon O'Connor, who was then the defence minister; deputy ministers Ward Elcock (Defence) and Peter Harder (Foreign Affairs); associate deputy minister William Elliott[5] (Public Security); Liberal continentalist Anne McLellan, Canada's former deputy prime minister and a defender of the oilpatch; the Alberta minister of energy, Greg Melchin; General Rick Hillier, Canada's chief of defence staff; former Conservative cabinet minister Perrin Beatty, now president of the Canadian Chamber of Commerce; the infamous continentalist Thomas d'Aquino, head of the Canadian Council of Chief Executives; Rear Admiral Roger Girouard; Major-General Daniel Gosselin; plus numerous top corporate heads, lawyers, petroleum industry officials, and others.

Among the many scheduled American participants were the political advisor to the head of the U.S. Northern Command; the president for the Americas of Lockheed Martin Corporation; the senior director for the Western Hemisphere of the American National Security Council; the U.S. deputy undersecretary of defense; Carla Hills, who was the primary U.S. NAFTA negotiator; the senior United States Air Force military assistant to the then secretary of defense, Donald Rumsfeld; the commander of U.S. Northern Command; the chair of the U.S. President's Council of Advisors on Science & Technology, Dr. James Schlesinger; the former American secretary of energy and defense; the deputy secretary of energy; plus many other top business, government, and military officials, and representatives of similar groups from Mexico.

In the intended-to-be-secret "internal agenda document" were plans for detailed discussions about economic, energy, security, military, and other forms of integration.

After I distributed the list of participants and the agenda to the media and to my e-mail list, many concerned people across Canada phoned the participants seeking more information. Almost no calls were returned, and those that were produced zero answers to the many questions asked about what was decided at the three-day meeting, who paid for the meetings, who organized them, and why every attempt was made to keep them secret.

Six months after the meetings, *Ottawa Citizen* reporter Kelly Patterson revealed that

> organizers of a controversial summit of top Canadian, US and Mexican politicians, military brass and business executives hired a consulting firm to keep the proceedings secret, access to information documents show.
>
> A "media management plan" for the event in Banff last fall imposed a gag order on all participants who were directed "to avoid direct media engagement where feasible."[6]

After I sent him the information I had about the meetings, Peter C. Newman wrote to me, "I tried phoning people I trusted and I thought trusted me, and no one would tell me anything. *IF* it isn't a conspiracy, they're doing their best to make it appear like one."[7]

In February 2007, Teresa Healy, senior researcher in social and economic policy at the Canadian Labour Congress, wrote about more closed-door meetings. The Security and Prosperity Partnership meetings seek "to avoid legislative change and public debate. Democratic debate and decision is making way for privileged corporate access and new rules that undermine sovereignty and human rights." Knowing that a new government treaty like the FTA or NAFTA for further Canada-U.S. integration would never survive the opposition in Parliament, "Proponents have moved underground to promote 'deep integration' . . . policy harmonization that increasingly opens social life across the continent to the discipline of the market. It is about increasing the power of corporations and ongoing de-regulation."

The so-called "NAFTA-plus" or "deep integration" or "Grand Bargain" being promoted and discussed at these secret meetings plans for the elimination of barriers to even greater foreign ownership and control, the slashing of government spending on social programs, the weakening of prospective environmental regulation (ensuring there are no barriers to increased energy and resource exports from Canada),[8] and many other policies that will result in even greater control of Canada by the United States. In other words, even more of the same kinds of policies you've been reading about in this book, but this time integration by the powerful via secrecy and stealth, planned by and for the likes of the Canadian Council of Chief Executives, the U.S. Council of the Americas, and the Center for Strategic Studies, clearly under the guidance and with the financing of the U.S. government in Washington, D.C.

Author Silver Donald Cameron says of the Security and Prosperity Partnership, "The SPP is the new name for the old American project of Manifest Destiny – absolute control over the whole continent." Cameron calls the CCCE the Canadian Council of Collaborators.

While our own government in Ottawa silently condones the plans for further integration with the United States, and while our provincial governments continue to be completely sound asleep on this vitally important topic, irony of ironies, at this writing, 15 U.S. states have expressed concern that the big-business-sponsored Security and Prosperity Partnership is a process that, wait for it, is a threat to States' Rights and to the sovereignty of the United States. And in July 2007, the U.S. House of Representatives cut some SPP funding because of the group's failure to consult Congress, complaining that "they have been intransigent, they have been unresponsive and frankly, they have been secretive."[9] In August 2007, 22 members of the House of Representatives asked George Bush to back away from the SPP because they had concerns that the SPP could undermine U.S. sovereignty and because they strongly objected to important discussions continuing "out of public view and without congressional oversight or approval" or the "proper transparency and accountability."

Does anyone think a single person in the entire Harper government has uttered a similar word of complaint? Not a chance.

On the surface, it may appear that not much happened at the follow-up August 2007 SPP meetings at Montebello, Quebec. But Thomas d'Aquino gives a hint about what is really happening behind the scenes: "A lot of work is going on, on the regulatory front, the environmental front, the energy front, the border front. . . . There is a big selling job that has to be done."[10] And according to him, "The Montebello Summit produced significant progress across a range of policy areas."[11]

Shortly after 9/11, d'Aquino had called for "more fundamental harmonization and integration with the U.S." to keep the borders open.

All of this covert, under-the-table planning is happening in Canada despite the fact that the overwhelming majority of Canadians, year after year, in poll after poll after poll, show that they believe that Canada has been too good a country for us to surrender to our greedy and selfish big-business leaders whose values are so very different from those of most Canadians who are still very proud of our country, our culture, our history, our values, our civility and tolerance, and are not the slightest bit interested in further "harmonization and integration" with the United States.

Alas, the compradors, the neo-cons, and the continentalists are powerful, well-organized, and very well-financed. We now know that the initial SPP agenda included over 300 items. In the words of a spokesman for the CCCE, "Many of these represent very small steps. . . . On the other hand, even 300 small steps, if we take them all, add up to a pretty giant leap for North America."

Testifying before the House of Commons Standing Committee on International Trade, Bruce Campbell, executive director of the Canadian Centre for Policy Alternatives, said of the SPP,

> I never hear talk about measures that would encourage upward harmonization of labour or environmental standards . . . tax measures that would prevent corporations from engaging in transfer pricing or discourage shifting profits to tax havens. This type of cooperation is not on the SPP agenda, and it begs the question: prosperity for whom?

Convergence and harmonization means . . . Canada bending its regulations or simply adopting U.S. federal regulations, and I ask the question: at what point does the narrowing of policy room to manoeuvre fundamentally compromise democratic accountability in our political system?[12]

In 2007, a group of well-known Canadians, including authors, academics, musicians, and former politicians, joined forces to oppose further integration with the United States. They issued a statement:

> Canada faces a stark choice. We can be gradually assimilated and lose our identity as a nation, or we can retain our independence and renew our own unique vision of what we wish for Canada's future . . . a vision of strong communities, tolerance, equality, environmental stewardship, and a peaceful and constructive role in the world.
>
> We believe that in spite of recent developments, Canadians believe passionately in the traditional values that guided this country in its post-war nation building.
>
> The Security and Prosperity Partnership will give us neither security nor prosperity, nor is it a genuine partnership. We stand against this scheme and urge other Canadians to join us.

In August 2007, the National Council of Women of Canada summed it all up nicely:

> This agreement has not been debated in or sanctioned by our Parliament. We believe that it threatens our sovereignty and puts control of our natural resources such as the Tar Sands and water at risk. For the U.S. government to negotiate a trade agreement manifestly to the advantage of international business interests using the "motherhood and apple pie" issue of security and prevention of terrorism is highly suspect. It's the

21st century version of "if you're not for us, you're against us." For a Canadian government to agree to such an undebated surrender of our sovereignty is shameful and unacceptable.

We demand that all SPP discussions be brought into the legislative and public domains immediately.

All of this said, there is one very important thing we have on our side: democracy. If we use it properly, we can overwhelm our own selfish sellouts. The important changes in election financing described earlier in this book are a huge step towards diminishing the influence of those who are now actively planning to give our country away. Now what is needed is for everyone who reads this book to go to work to enlist a great many others to help us preserve and improve the country we so love.

There's no doubt in my mind that we can stop these sellouts. But for us to do so requires nothing less than a very substantially heightened degree of direct political activism by people who really care about the survival of our country. If you – and I mean you – don't decide to do this, and do it soon, it is a certainty that it will quickly be too late for your children and your grandchildren to do so in the future. Continentalist conservative *Globe and Mail* columnist Neil Reynolds wrote on January 2, 2008, "For those folks who still resist the economic integration of North America, and there are a strident few left, it's getting awfully late for a comeback."

You've already seen in this book how profoundly Canada has changed as a result of the policies promoted by the same people who are now secretly planning to change our country even more. If you care about Canada, now is the time to go to work to preserve our sovereignty, our independence, our cherished and special values, and our economic, social, and political integrity. Don't be misled by intentionally leaked reports out of Washington that the SPP is dead. The likes of Tom d'Aquino, Raymond Chrétien, John Manley, Peter Lougheed, and Allan Gotlieb are powerful, persistent, tenacious, and well-connected in both countries. At this writing, there are at least 20 cross-border big-business committees working on integration.

Let's make sure that Canadians know of and understand their plans, and let's be certain that we never allow them to succeed.

Finally, as the manuscript for this book goes out for typesetting, there is again much talk about yet another federal election. It's likely that Harper, despite what he says to the contrary, has wanted one when he has been comfortably ahead in the polls, while Dion, despite his bluster, and most Liberals and most Canadians don't. Harper has all the advantages of incumbency, a huge war chest of election funds, and polls that show Canadians would much prefer him to Mr. Dion as prime minister. Meanwhile, while Dion is widely regarded as a very nice man, whatever the antonym for charisma is, Dion has it. His performance as Liberal leader has been uninspiring. Party fundraising is poor, and it's no secret that by the fall of 2007 some Liberals had already been trying to think up ways of replacing him.

By late 2007, an Angus Reid poll said Stephen Harper was preferred as Prime Minister by 33 percent compared to only 14 percent for Stéphane Dion. Of interest is that the majority of those polled opted for neither.

Everyone knows that public opinion can change dramatically before and during an election campaign, but keep in mind that there will be a powerful, sustained Conservative television campaign that will likely prove effective. Most pundits predict another minority Conservative government. But if Stephen Harper does win a majority, Canada will almost certainly change for the worse, even more than all the changes already described in this book put together.

* * * *

A great deal has happened since the October 2008 federal election. Needless to say, millions of Canadians were very relieved that Stephen Harper did not win a majority government. Nevertheless, the prospects are grim. If you would like to read my comments on events as they unfold, please go to vivelecanada.ca and follow the link to Mel Hurtig.

GLOSSARY

BCNI – Business Council on National Issues (see CCCE)

Campaign 2000 – a non-partisan, cross-Canada coalition of over 120 national, provincial, and community organizations committed to working together to end child and family poverty in Canada

CCCE – Canadian Council of Chief Executives. This big-business lobbying organization is composed of 150 of the largest, most influential corporations in Canada, both Canadian and foreign. Previously named the Business Council on National Issues (BCNI), it changed its name to the Canadian Council of Chief Executives in December 2001. The CCCE's chairman is Gordon M. Nixon, CEO of the Royal Bank of Canada. Its chief executive and president is Thomas d'Aquino, Canada's leading corporate advocate of deeper integration with the United States. Collectively, the CCCE administers some $3.2-trillion in assets and has annual revenue of some $750-billion. The CCCE is almost certainly the most continentalist corporate organization in modern Canadian history. The BCNI was the leading and most powerful proponent of both the FTA and NAFTA.

CCPA – Canadian Centre for Policy Alternatives. This is an independent research institute concerned with issues of social and economic justice. Founded in 1980, the CCPA is one of Canada's leading progressive voices in public policy debates. The CCPA is a registered, non-profit organization supported by more than 10,000 members across Canada, with a national office in Ottawa and provincial offices in Nova Scotia, Ontario, Manitoba, Saskatchewan, and British Columbia.

CIHI – Canadian Institute for Health Information

comprador – agent of a foreign power

continentalist – a term used to describe Canadians who believe in greater integration with the United States

direct investment – Direct investment refers to the acquisition of what is intended to be a lasting, influential financial interest in an enterprise. It usually refers to investment of a resident entity in one country obtaining a long-term interest in an enterprise in another nation.

EU – European Union, presently composed of 27 countries, with possible future expansion

EU15 – The member countries in the European Union prior to the EU's expansion in May 2004. The countries of the EU15 are: Austria, Belgium, Denmark, Finland, France, Germany, Greece, Ireland, Italy, Luxembourg, the Netherlands, Portugal, Spain, Sweden, and the United Kingdom.

FDIIC – foreign direct investment in Canada. Direct investment usually represents significant ownership and control of a corporation, although in some widely held corporations as little as 10 percent ownership results in effective control.

FIRA – Foreign Investment Review Agency

FTA – Canada-U.S. free-trade agreement, which came into effect in 1989

G7, G8 – Group of Seven, Group of Eight. The G7 is a group of seven leading industrialized countries: Canada, France, Germany, Britain, Italy, Japan, and the United States. The G8 includes Russia.

GATT – General Agreement on Tariffs and Trade

GDP – gross domestic product. The total output of goods and services for final use produced by both residents and non-residents in an economy. This is the standard measure of the incomes generated from productive activity in a country.

GNI – gross national income. GNI is equal to GDP plus the net receipts of wages, salaries, and property income from abroad.

IMF – International Monetary Fund

labour productivity – This key economic indicator is both important and confusing. For some, a poor performance in labour productivity implies workers are not working long enough or hard enough. But for others, it means the workers aren't being given the updated technology and equipment they need to be more competitive. This last definition can be directly tied to poor corporate investment in research and in machinery and equipment.

For Statistics Canada, labour productivity measures real gross domestic product per hour worked and is a primary determinant of improvements to the standard of living in the long run and also the main source of economic growth.

Labour productivity is defined as the real output per hours worked. Multifactor productivity measures the efficiency with which inputs are used in production.

LDC – least developed countries. The group of 50 countries classified by the United Nations as falling below thresholds established for income, economic diversification, and social development.

LICOs – Low-income cut-offs are Statistics Canada thresholds, which are determined by analyzing family expenditure data, below which families will likely devote a larger share of income to food, shelter, and clothing than would the average family, adjusted for family size and community. Statistics Canada says LICOs are "a well-defined methodology that identifies those who are substantially worse off than the average" and those who are in "the most straitened circumstances."

low income – Low income is broadly defined by Statistics Canada as a family or individual income that is half or less of median income.

market income – Market income is the total of earnings from employment, invested income, and retirement income. It is the same as total income less government transfers.

median – The median is the point where one half of incomes (for example) is higher and one half is lower.

multi-factor productivity – This is where labour, capital, technological progress, and management considerations are added together to measure growth. Canada does better in this measurement than it does when only labour productivity is considered. From 1995 to 2003 inclusive, of 19 OECD countries, Canada was about in the middle of the pack with the ninth best record. Ireland, Finland, Greece, Australia, the United States, France, Sweden, and the United Kingdom all outperformed Canada, but Canada bettered Portugal, Germany, New Zealand, Austria, Belgium, Japan, Spain, the Netherlands, Denmark, and Italy, which had by far the worst record.[1]

NAFTA – North American free-trade agreement. The Canada-U.S.-Mexico agreement that came into effect in 1994.

National Child Benefit Supplement – The National Child Benefit was launched in 1998 by the federal, provincial, and territorial governments to increase child benefits to low-income families with children. The National Child Benefit Supplement is the federal government's increased contribution to the initiative.

OECD – Organization for Economic Cooperation and Development. This is composed of 30 developed countries which work together on economic, social, globalization, and environment matters. The OECD publishes a steady stream of invaluable books, papers, articles, and studies. The OECD member countries are Australia, Austria, Belgium, Canada, the Czech Republic, Denmark, Finland, France, Germany, Greece, Hungary, Iceland, Ireland, Italy, Japan, Korea, Luxembourg, Mexico, the Netherlands, New Zealand, Norway, Poland, Portugal, the Slovak Republic, Spain, Sweden, Switzerland, Turkey, the United Kingdom, and the United States.

portfolio investment – The purchase of shares and bonds for income yields or capital gains, not with the objective of asserting ownership or control.

poverty – There is no official measurement of poverty in Canada. Social groups use the term to refer to those with incomes below the Statistics Canada LICO lines. Internationally, poverty is usually defined as living with an income less than 50 percent of the median disposable income. In common usage, poverty means being deprived of the necessities of life, suffering, hunger, lack of proper shelter and clothing, and the inability to take part in the community in a reasonable manner. (See Appendix Two for various methods of measuring poverty.)

purchasing power parity (PPP) – Purchasing power parities are the currency conversions used to eliminate differences in price levels between countries. They provide a system for comparing the volume of GDP in different countries. PPPs are determined using the cost of a basket of goods and services in different countries. They are calculated in national currency units per U.S. dollar.

R&D – research and development

transfer payments – Transfer payments generally refer to payments by Ottawa to the provinces or to individuals, or other payments made by the provinces. Transfer payments to individuals include old age security, Canada and Quebec pension plans, employment insurance, guaranteed income supplements, spouse's allowances, child benefit payments, welfare, workers' compensation benefits, GST credits, etc.

United Nations Human Development Index (UNHDI) – This is based on longevity, adult literacy, school enrolment, and real GDP per capita and is published annually.

Unicef – United Nations Children's Fund

APPENDIX ONE: HEALTH CARE

CATEGORY BREAKDOWN OF TOTAL HEALTH-CARE EXPENDITURES IN CANADA

The Canadian Institute for Health Information (CIHI) says that in 2006, 29.8 percent of health-care funds went to hospitals, 17.0 percent for drugs, 13.1 percent to physicians, 9.4 percent to other institutions, 5.8 percent for public health costs, 4.1 percent for capital expenditures, 3.9 percent for administration, and the balance for "other health spending."

Of the $39.24-billion in private sector health expenditures in 2004, 48.2 percent came from personal "out-of-pocket" spending, 41.3 percent was from private health insurance, and 10.5 percent from other sources. Most private sector health spending is for prescribed drugs, dental care, vision care, hospital accommodation, over-the-counter drugs, non-hospital institutions such as nursing homes and residential care facilities, and personal health supplies. In 2006, about 61.8 percent of all prescribed drugs had to be privately financed.

It should be noted that in the mid-1980s, spending for physician services was up to 15.7 percent of total health spending, but in 2006 it was down to 13.1 percent. Mind you, there were additional expenses for physicians employed directly by hospitals, boards of health, etc.

"Other professional" expenses amounted to over $15-billion in 2006. These include the costs of dentists, denturists, optometrists, opticians, chiropractors, physiotherapists, and private duty nurses.

PROVINCIAL AND TERRITORIAL HEALTH-CARE SPENDING

The CIHI has forecast that provincial and territorial health spending in 2006/2007 would be $96-billion, up over $5-billion from the previous year and up almost $11-billion from 2004/2005. (After inflation, the 2006/2007 amount reflects real growth of 3.4 percent over 2005/2006.) Total provincial and territorial government health spending in 2006/2007 was expected to represent some 39 percent of all provincial and territorial government spending.

In 2007, per-capita spending in Alberta was $5,390, about $1,000 more than in Quebec.

In 1975, provincial and territorial spending amounted to 71.4 percent of total health-care expenditures. But by 2004 this had fallen to 64 percent.

PUBLIC AND PRIVATE HEALTH-CARE COSTS

The October 2006 edition of the *OECD Health Data* report says that in 2004 private per-capita health expenditures in the United States were $3,375, followed by Switzerland at $1,695, the Netherlands at $1,144, Australia at $1,014, and Canada in fifth place at $955 (all figures in U.S. dollars).

In 2004, private health-care expenditures in the United States were 8.5 percent of GDP, compared to about 3.0 percent in Canada.

Many Canadians think that the public share of total health expenditures in Canada is higher than in most countries. Not so. In 2004, Canada, at 70 percent public, was actually slightly below the 73 percent OECD average. All the following countries were above 80 percent in their publicly funded share of their total health spending: the Czech Republic and Luxembourg, both at 90 percent; the Slovak Republic at 88 percent; Sweden at 85 percent; Norway and Ireland at 84 percent; the United Kingdom and Denmark at 83 percent; and Japan at 82 percent. Only Switzerland at 59 percent, Greece at 51 percent, Korea at 49 percent, Mexico at 46 percent, and the United States were below 60 percent in their public share of health expenditures.

In an OECD list of 21 countries, in 2004 France led the way in public health expenditures as a percentage of GDP at 8.3 percent, followed by Germany and Norway at 8.1 percent. Canada was in eighth place at 6.9 percent.

If we measure public per-capita health expenditures in U.S. dollars for 2004, Luxembourg came in at $4,603, followed by Norway at $3,311. Canada was down in seventh place at $2,210.

HEALTH CARE IN THE UNITED STATES

Speaking in Scranton, Pennsylvania, in January 2002, George W. Bush said, "I'm proud of our health-care system." Yet in 2007, a Kaiser Family Foundation poll said that Americans worried about their health care more than losing their job or being attacked by terrorists.

According to the World Health Organization, the United States ranked 37th in overall health performance and 54th in the fairness of the health-care system. International estimates are that in 2006 the United States was, on average, in 12th place out of the top 13 industrialized nations in a list of 16 key health indicators. And in a list of 17 countries, 14 had a higher percentage of patient satisfaction than in the United States.

American women are 70 percent more likely to die in childbirth than women in Europe.

The infant mortality rate in the home of the U.S. capital, the District of Columbia, is twice as high as in Beijing. In 2002, the number of babies who died in D.C. before their first birthday was 11.5 per 1,000 compared with 4.6 in Beijing. In the number of deaths among children under the age of five per 1,000 live births, the United States is down in 26th place, with a rate 75 percent higher than the rates for the Czech Republic, Finland, Italy, Japan, Norway, Slovenia, and Sweden.

From a featured article in the *Boston Globe*:

MY FELLOW AMERICANS: WANT A HEALTH TIP? MOVE TO CANADA

An impressive array of comparative data show that Canadians live longer and healthier lives than we do. What's more, they pay roughly half as much per capita as we do for the privilege.[1]

A study of health disparities in the United States and Canada was published in the July 2006 edition of the *American Journal of Public Health*. Among its findings were that Canadians had lower rates of diabetes, asthma, hypertension, arthritis, heart disease, and depression. Canadians reported that they were unable to afford their needed medicines at half the rate of Americans.

In 2005, the U.S. *Archives of Internal Medicine* reported that heart bypass surgery in Canada costs an average of $10,373, compared to $20,673 in the United States (both in U.S. dollars). The same report shows drug costs, lab tests, hospital costs, and other medical costs were much higher in the United States than in Canada.[2] Nine percent of Canadians reported that they sometimes did not fill prescriptions because of costs, compared to 21 percent of Americans.

In May 2006, a Harvard Medical School study found that

> Americans, who spend twice as much per capita on health care than Canadians, had higher rates of nearly every serious chronic disease examined in the survey including heart disease, diabetes, arthritis and asthma.
>
> Canada, despite spending so much less, is actually getting better outcomes in terms of both health and access to care.
>
> Dr. Steffie Woolhandler and her peers concluded that Canada's public health system was one of the key reasons why Canadians fared better.

Economist Paul Krugman says that if the United States had the same type of single-payer health-care system as Canada, the savings would be more than $200-billion a year, much more than the cost of providing insurance for every one of the many millions of Americans with no insurance. By the spring of 2006, administration of health-care costs in the

United States came to about 25 cents of each dollar, compared to two cents of every dollar in Canada.

Gerard F. Anderson, professor at the Johns Hopkins Bloomberg School of Public Health, complaining about the failure of the United States to switch to a single-payer health-care system, says, "The story never changes. The United States is twice as expensive with about the same outcome. As a consumer, I don't mind paying more if I'm getting more, but that's just not the case in the U.S."

In 2005, the Lewis Group, a U.S. health-care consulting firm, said that universal health coverage in California would have saved $8-billion in 2006, and from 2006 to 2015 estimated savings would be $343-billion.[3]

In 2005, the prestigious *American Journal of Medicine* reported that the quality of care for patients is better when care is provided through not-for-profit health-care plans than through for-profit plans. Another report from the Harvard School of Public Health "reinforced the concern that the financial incentives of for-profit plans lead to less aggressive efforts to manage the quality of care."[4]

By the end of 2006, the number of Americans without health insurance had grown for the sixth year in a row, increasing to 47 million, some 14 million greater than the entire population of Canada. The number of people in the United States with employer-provided health insurance fell by more than three million in 2005, to only 56 percent of employees. Some 8.3 million children, about 11.2 percent, had no health insurance.

In 2003, over 43 percent of the non-elderly population in Texas had no health insurance. In California it was over 37 percent, and in Florida 35 percent. Associated Press reported that in 2004 "nearly 82 million people – one third of the U.S. population younger than 65 – lacked health care insurance at some point over the past two years."[5] Of the 47 million Americans with no health insurance, 82 percent live in households where the head of the family had a job. More than one-third of the uninsured have family incomes of $40,000 or more. According to a *New York Times* editorial (August 26, 2007), "Families with incomes of $70,000 or more can be hard pressed to pay for private insurance in high cost areas."

The percentage of children not covered by employer-provided health insurance has increased for six years in a row and has now risen to 40 percent.

In 2004, a shocking 40 percent of American adults surveyed said that they have had to go without needed health care due to costs. For those with below-average incomes, the percentage increases to 57 percent.[6]

In the words of Paul Krugman, writing in the *New York Times,* employer health insurance in the United States is unravelling with a "crazy quilt of private insurers" in an "extremely inefficient health care system" that now costs the average employer $10,000 for a family of four, and private health insurance for a family of four now costs an average of some $12,000 a year. According to Krugman, "the U.S. is now spending more on health care, but the results are worse."[7]

Some 19,000 American men and women die every year directly because they had no health insurance. Where in 1993 some 60 percent of American employers had company health-care plans, by 2005 that number was down to 45 percent and rapidly decreasing.

By 2005, the percentage of Americans who could afford private health-care insurance was at its lowest level in 19 years. While over 86 percent of Americans in the highest income quintile had insurance, fewer than 22 percent in the lowest quintile were insured.

In the first five years of the George W. Bush presidency, the cost of health insurance premiums rose 73 percent. One projection is that by 2008 the average annual premium for employer-sponsored family coverage would be over $14,500, more than double the cost in 2001. Many employer health insurance plans have deductibles ranging from $1,000 to $5,000, and a choice of a doctor or hospital is not often possible.

By 2013, predictions are that 56 million Americans will have no health-care insurance.

In 2005, over 700,000 households declared bankruptcy in the United States because they were unable to pay health-care bills. Over two million Americans were directly affected in this way. Yet almost 70 percent of these people had some form of health insurance.

Paul Krugman writes, "The health-care crisis in the U.S. is back, both because medical costs are rising rapidly and because we're living in an increasingly Wal-Martized economy, in which even big, highly profitable employers offer minimal benefits."[8]

In March 2007, the *New York Times* reported that a majority of Americans said that the federal government should guarantee health insurance to every American, especially children, and that they are willing to pay higher taxes for this.

In October 2007, government audits showed that tens of thousands of U.S. Medicare recipients had claims improperly denied by private insurers, including people with HIV and AIDS. Many of the private insurers were accused of not answering phone calls from patients, doctors, and drugstores.

SOME SAMPLE AMERICAN SURGICAL COSTS IN U.S. DOLLARS

Angioplasty	$57,262 to $82,711
Simple heart valve replacement	$159,325 to $230,138
Hip replacement	$43,780 to $63,238
Knee replacement	$40,640 to $58,702[9]

APPENDIX TWO: POVERTY

MEASURING POVERTY

Statistics Canada has never had an official poverty line. It does, however, measure the number of Canadian who they define as living on "low incomes." In 2004, some 3.5 million Canadians were living on low incomes. In recent years, the number peaked in 1996 at about 4.6 million.[1] However, some other measurements place the number as high as 5.7 million.[2]

Some of our extreme right-wing, anti-poor fanatics object to the idea that the Statistics Canada low-income cut-off is a reliable definition of a proper poverty line. Witness, for example, the regular attacks on the annual reports of Campaign 2000 by the *National Post*,[3] Montreal economist William Watson, the far-right Fraser Institute, and Chris Sarlo, among others.

But as Laurel Rothman, national coordinator of Campaign 2000, has pointed out in answer to the critics of low-income cut-off measurements that have been used in Canada for over two decades, the newer market basket measure commissioned by the federal, provincial, and territorial governments as "a consensus definition of basic needs/poverty levels" after years of study produces results similar to the low-income measurements, but with an even higher child poverty rate of 16.9 percent compared to 16.6 percent. (Remember that in 1989, when the House of Commons promised to abolish child poverty by the year 2000, the rate was 15.1 percent.)

Campaign 2000's November 2007 report about poverty in Canada shows that despite our growing economy, in 2005, using after-tax measures, the rate of child poverty was 11.7 percent, exactly what it was sixteen years earlier in 1989. Using the before-tax measure commonly employed until recently, child poverty increased from 15.1 percent to 16.8 percent.

In its 2007 report, Campaign 2000 uses both before-tax and after-tax figures. For obvious reasons, government and war-against-the-poor commentators much prefer the after-tax numbers which are much lower, since they include both taxes and government transfers.

Michael Goldberg of Vancouver, a member of the Campaign 2000 National Steering Committee explains further:

> Most of the public reports by Statistics Canada now use the after-tax figures. It seemed to us that arguinig over what measures should be used took us away from the task at hand, namely getting governments to commit to a poverty reduction plan that would eliminate poverty in Canada.

If you prefer the lower after-tax figure, that still leaves a disgraceful 788,000 Canadian children living in poverty, roughly one in eight.

In fact, all three commonly accepted measurements of poverty in Canada have produced similar results, yet the federal government has refused to recognize any of them officially despite the fact that the statistics are comprehensive and reliable. So, year after year goes by without any official recognition of Canada's disgraceful poverty figures.

Could it be that both Liberal and Conservative governments have been too embarrassed by the shameful annual survey results, especially when they are compared with those of so many other developed countries?

SOME NOTES ABOUT POVERTY IN THE UNITED STATES

As indicated earlier, the United States has by far the highest rates of poverty in the industrialized world, with some 22 percent of all children living in poverty. And as I have pointed out often in the past, the official

U.S. method of measuring poverty, essentially unchanged since 1963, is regarded almost everywhere as a sad joke. In 2005, the U.S. rate was said to be 12.7 percent, but according to much more reliable international estimates, the real rate is far above that figure. The renowned Luxembourg Income Study puts the American child poverty rate at about 22 percent, compared to the Canadian rate of 16.6 percent.

As low as the *official* American poverty figures are compared to reality, even the official rates are horrendous in some parts of the country: almost 34 percent in Detroit, just under 25 percent in Philadelphia, almost 29 percent in Miami, and over 20 percent in New York City. This despite the fact that over 40 years ago, President Lyndon Johnson declared a "war on poverty" in his 1964 State of the Union address.

Incredibly, an American family with two children and a total income of $19,850 ($1,650 a month) is not considered poor even if they live in a city where average rent in a working class neighbourhood is well over $1,000 a month.

Unicef explains: "The official U.S. poverty line dates back to concepts and judgements made in the 1960s. . . . In August 2000, 40 prominent scholars sent an open letter to senior government officials stating that unless 'we correct the critical flaws in the existing measure, the nation will continue to rely on a defective yardstick.'"[4]

And why has the official measurement of poverty in the United States been left at such a ludicrously unrealistic level? The *New York Times* explains: "During the last four decades, no president of either party has wanted to draw attention to a statistic that the nation has come to take for granted, especially if updating it might cause the number of people regarded as living in poverty to increase."[5]

And what do most Americans think about the millions of poor men, women, and children? A Pew Research Center poll in 2005 showed that between 40 and 50 percent of Americans say that "poor people have it easy."

And of course this attitude is reflected in public policy. In the mid-1990s, some 4.4 million poor American families received welfare payments. In 2006, that number was down to only 1.9 million families.

In 2006, almost six million people were living in poverty in California, where an average family in the lowest 10 percent of incomes had an income of some $15,600, a decline of 12 percent since 1969. A very modest two-bedroom apartment in San Francisco rents for $21,300 a year.

It's interesting to compare the disposable incomes of the poor in the United States and Canada. In Canada, the poorest 10 percent have almost 60 percent more disposable income than the poorest 10 percent of Americans. The next 10 percent have almost 20 percent more disposable income.

The Pew Research Center says two in five low-income Americans at times go hungry, the highest such proportion in the industrialized world, compared to one in four low-income Canadians and Britons according to the well-respected research centre.

In 1974/1975, 13 percent of Americans said they periodically couldn't afford to see a doctor. By 2002, the percentage had grown to 26 percent. As many as 55 percent of low-income Americans occasionally could not afford to pay for care, while only 17 percent of low-income Germans and 8 percent of low-income Japanese faced the same problem.

A February 2007 report said that

> the percentage of poor Americans who are living in severe poverty has reached a 32-year high, millions of working Americans are falling closer to the poverty line, and the gulf between the nation's "haves" and "have-nots" continues to widen.
>
> Sixteen million Americans are living in deep or severe poverty as wages are dwarfed by profits.
>
> Household income for working age families, adjusted for inflation, has fallen for five straight years.
>
> Forty-three percent of the nation's poor are living in deep poverty, the highest rate since at least 1975.[6]

NOTES

PART ONE

CHAPTER ONE

1. *Human Development Report,* 2006. United Nations, for period 1990 to 2004.
2. E-mail to the author, March 21, 2006.
3. *Vancouver Sun,* July 29, 2006.
4. *OECD Health Data,* 2007.
5. Dr. Richard Levin, dean of medicine, McGill University, *Globe and Mail,* April 18, 2007.
6. *Health at a Glance,* 2005, OECD.
7. *Globe and Mail,* July 31, 2007.
8. *Globe and Mail,* August 1, 2007.
9. See www.healthcoalition.ca.
10. *Toronto Star,* August 15, 2007.
11. *Human Development Report,* 2005. United Nations.
12. *OECD Factbook,* 2006.
13. *The Economist* magazine's *Pocket World in Figures,* 2007.
14. January 13, 2007.
15. Unicef *Innocenti Report Card* 7, United Nations, February 2007.
16. *OECD Factbook,* 2004.
17. *The State of the World's Children,* 2006. United Nations.
18. Ibid.
19. *OECD Observer,* October 2006.
20. *Health at a Glance,* 2005, OECD.
21. *The State of the World's Children,* 2006. United Nations.
22. *Health at a Glance,* 2005, OECD.
23. Ibid.
24. Ibid.
25. Canadian Institute for Health Information.
26. *Health at a Glance,* 2005, OECD.

27. CanWest News Service, July 10, 2007, and the Canadian Centre for Substance Abuse, November 24, 2004.

28. *Health at a Glance*, 2003, OECD

29. *Globe and Mail*, April 19, 2007.

30. Testimony to the Canadian Senate's Kirby committee hearings of the Standing Committee on Social Affairs, Science and Technology.

31. The most recent comparative figures show that 25 of the 30 OECD countries have a larger number of practising physicians per 100,000 population than Canada.

CHAPTER TWO

1. Campaign 2000 is a cross-Canada public education movement with 120 national, community, and provincial partners actively involved in its work.

2. The B.C. child poverty rate is the highest in Canada.

3. *Toronto Star*, January 20, 2006.

4. *Toronto Star*, July 30, 2007.

5. *Hunger Count 2006*, Canadian Association of Food Banks.

PART TWO

CHAPTER THREE

1. Louise Brown, *Toronto Star*, November 24, 2006.

2. See the work of Howard Sapers, Canada's prison ombudsman.

3. Of the about 652,000 aboriginal people age 15 or over, some 61 percent live in Western Canada.

4. *The Daily*, Statistics Canada, June 6, 2006.

5. *Globe and Mail*, March 5, 2007.

6. *The Daily*, Statistics Canada, January 29, 2007.

7. *Globe and Mail*, March 21, 2007.

CHAPTER FOUR

1. *OECD Factbook*, 2007.

2. For a more detailed look at comparative levels of program spending, see the chapter on government.

3. Vincent Calderhead, *Toronto Star*, May 25, 2006.

4. *The Social Benefits and Economic Costs of Taxation*, 2007.

CHAPTER FIVE

1. OECD *Database on Full-time/Part-time Employment*, Paris, 2005, and *OECD Factbook*, 2007.

2. OECD *Employment Outlook*, June 2005, and *Annual Labour Force Statistics*, 1984–2004, OECD, 2005.

3. *Globe and Mail*, June 9, 2007.

4. *Canadian Economic Observer*, Historical Statistical Supplement, 2005/06, Statistics Canada.

5. *The Economist* magazine's *Pocket World in Figures*, 2006.

6. *OECD Economic Outlook*, June 2007.

7. *The Economist* magazine's *Pocket World in Figures*, 2006.

8. June 4, 2006.

CHAPTER SIX

1. *Toronto Star*, May 16, 2006.

2. *Society at a Glance*, 2006, OECD.

CHAPTER SEVEN

1. August 28, 2006.

2. *Toronto Star*, August 23, 2006.

3. *Policy Options*, April/May 2006.

4. Colin Hughes, *Toronto Star*, November 2, 2006.

5. Carol Goar, *Toronto Star*, October 11, 2006.

6. May 9, 2006.

7. Thomas Walkom, *Toronto Star*, September 9, 2005.

8. December 15, 2006.

CHAPTER EIGHT

1. *OECD Factbook*, 2006.

2. Ibid.

3. *OECD Factbook*, 2007.

4. *The Economist,* November 18, 2006.

5. James Travers, *Toronto Star,* June 8, 2005.

6. *New York Times,* September 28, 2005.

7. *The Daily,* Statistics Canada, January 30, 2007.

8. *The Daily,* Statistics Canada, September 10, 2007.

9. CanWest News Service, June 18, 2007.

10. September 27, 2005.

PART THREE

CHAPTER NINE

1. *Toronto Star,* March 18, 2005.

2. *Canadian Economic Observer,* Historical Statistical Supplement, 2005/06, Statistics Canada.

3. David Crane, *Toronto Star,* February 21, 2004.

4. *Poverty and Policy in Canada* (Toronto: Canadian Scholars' Press, 2007).

5. *Toronto Star,* January 12, 2005.

6. May 26, 2006.

7. *Perspectives on Labour and Income,* Statistics Canada, Winter 2006.

8. *Report on Business,* July/August 2006.

9. Janet McFarland, *Globe and Mail,* October 28, 2005.

10. *Promotion & Education,* Vol. XIII, No. 4, 2006.

CHAPTER TEN

1. *The Daily,* Statistics Canada, March 20, 2006.

2. "The Evolution of High Incomes in Northern America: Lessons from Canadian Evidence," *American Economic Review,* Vol. 95, No. 3, June 2005.

3. *Income Trends in Canada, 1980–2005,* Statistics Canada.

4. *Perspectives on Labour and Income,* Statistics Canada, Spring 2005.

5. "A Cautionary Note on Relying on Human Capital Policy to Meet Redistribution Goals," *Canadian Public Policy,* Vol. 33, No. 4, December 2007.

6. *Revisiting Recent Trends in Canadian After-Tax Income Inequality Using Census Data,* 2006.

7. "Pulling Apart: The Growing Gulfs in Canadian Society," *Policy Options,* April/May 2006.

8. *Globe and Mail,* January 5, 2006.

9. *Growing Gap, Growing Concerns.*

10. *Toronto Star,* May 7, 2007.

11. *Toronto Star,* March 1, 2007.

CHAPTER ELEVEN

1. René Morissette and Xuelin Zhang, "Revisiting Wealth Inequality," *Perspectives,* Statistics Canada, December 2006.

2. *The Wealth of Canadians: An Overview of the Results of the Survey of Financial Security,* 2005. Statistics Canada.

3. See *The Daily,* Statistics Canada, June 22, 2007.

4. *Globe and Mail,* December, 2006.

CHAPTER TWELVE

1. *Canadian Economic Observer,* Historical Statistical Supplement, 2006/07, Statistics Canada, p. 115.

2. *Report on Business,* July/August 2007.

PART FOUR

CHAPTER THIRTEEN

1. Report by TD Bank Financial Group chief economist Don Drummond, September 2007.

2. *Toronto Star,* July 16, 2007.

3. A. Tomas, *Canadian Economic Observer,* Statistics Canada, April 2006.

4. *OECD Economic Outlook,* December 2006.

5. *Globe and Mail,* April 19, 2007.

6. *Maclean's,* March 19, 2007.

7. *Financial Post,* January 6, 2007.

8. Allan Tomas, *Recent Trends in Corporate Finance,* Statistics Canada, March 2006.

9. *OECD Factbook,* 2007.

CHAPTER FOURTEEN

1. *Human Development Report,* 2005. United Nations.
2. *OECD in Figures,* 2005 supplement.
3. David Crane, *Toronto Star,* October 22, 2006.
4. *OECD Factbook,* 2007.
5. *Perspectives on Labour and Income,* Statistics Canada, Spring 2007.

CHAPTER FIFTEEN

1. *Globe and Mail,* June 29, 2005.
2. Ibid.
3. *Canadian Economic Observer,* Statistics Canada, September 2005.
4. *OECD Factbook,* 2007.
5. *The Daily,* Statistics Canada, March 26, 2007.
6. *Globe and Mail,* January 3, 2007.
7. Rao, Sharpe and Tang, "Productivity Growth in Service Industries: A Canadian Success Story." *Micro,* Winter/Spring 2005.

CHAPTER SIXTEEN

1. *The Daily,* Statistics Canada, January 5, 2007, and the seasonally adjusted Statistics Canada *Labour Force Survey.*
2. *The Economist,* October 1, 2005.
3. May 22, 2007.

CHAPTER SEVENTEEN

1. *The Daily,* Statistics Canada, February 14, 2007.
2. The deficit calculation is based on the industry definition, the North American Industrial Classification System.
3. Jim Stanford, *Canada's Deteriorating Automotive Trade Performance,* Canadian Auto Workers, October 2007.
4. *Financial Post,* February 15, 2007.
5. *The Daily,* Statistics Canada, May 17, 2007.
6. *National Post,* August 8, 2005.
7. *Globe and Mail,* September 25, 2005.
8. *OECD Factbook,* 2006.

CHAPTER EIGHTEEN

1. *OECD Factbook,* 2007.
2. *OECD Economic Outlook,* December 2006.
3. *Maclean's,* March 19, 2007.
4. June 1, 2007.
5. *National Post,* July 26, 2007.
6. *Globe and Mail,* October 12, 2006.
7. *Globe and Mail,* September 28, 2006.
8. *Globe and Mail,* October 31, 2007.
9. *National Post,* October 31, 2007.
10. *OECD Economic Outlook,* June, 2007.
11. *Globe and Mail,* November 2, 2007.
12. *Globe and Mail,* January 3, 2005.
13. *Toronto Star,* November 3, 2007.
14. *Globe and Mail,* October 31, 2007.
15. *Financial Post,* October 10, 2007.

CHAPTER NINETEEN

1. *Fiscal Reference Tables,* Department of Finance, September 2007.
2. *OECD Factbook,* 2007.
3. *Canadian Economic Observer,* Statistics Canada, May 2005.
4. *The Daily,* Statistics Canada, September 24, 2007.
5. *OECD Factbook,* 2004.
6. Ibid.
7. *The Daily,* Statistics Canada, March 14, 2005.
8. *Perspectives,* Statistics Canada, Autumn 2006.
9. Ibid.
10. *OECD Factbook,* 2006.
11. *Toronto Star,* May 30, 2004.
12. *Perspectives,* Statistics Canada, Autumn 2006.
13. *The Daily,* Statistics Canada, March 30, 2006.

CHAPTER TWENTY

1. *Globe and Mail,* September 10, 2006.

2. Third in natural gas production, eighth in copper, sixth in lead, fourth in zinc, second in nickel, third in aluminum, eighth in gold, and sixth in energy, according to the 2007 edition of *The Economist* magazine's *Pocket World in Figures.*

3. See Robin Banerjee and William B.P. Robinson, "Give Workers Better Equipment," *Toronto Star,* June 12, 2007, and the earlier *Globe and Mail* editorial on their work, May 21, 2007.

PART FIVE

CHAPTER TWENTY-ONE

1. *Education at a Glance,* 2007, OECD.

2. *Toronto Star,* November 4, 2007.

3. *The Daily,* Statistics Canada, November 7, 2006.

4. *OECD Factbook,* 2006.

5. *Education at a Glance,* 2006, OECD, and *The Daily,* Statistics Canada, November 7, 2006.

6. *The Daily,* Statistics Canada, November 7, 2006.

7. *Maclean's,* November 19, 2007.

8. *Globe and Mail,* September 2, 2006.

9. *The Daily,* Statistics Canada, February 8, 2007.

10. Tom Axworthy, *Toronto Star,* November 4, 2007.

11. Shirley Won, *Globe and Mail,* May 23, 2007.

12. *Toronto Star,* November 4, 2007.

13. *Toronto Star,* September 24, 2006.

14. *Education at a Glance,* 2006, OECD.

15. *The Economist,* June 11, 2005.

16. Statistics Canada's Centre for Education Statistics.

CHAPTER TWENTY-TWO

1. February 1, 2005.

2. February 4, 2007.

3. Data from Hill Strategies, quoted in the *Globe and Mail*, February 21, 2007.

4. *OECD Factbook*, 2007.

5. *The Economist* magazine's *Pocket World in Figures*, 2006.

6. *The Daily*, Statistics Canada, June 25, 2007.

7. Statistics Canada, *General Social Survey*, 2005.

8. *The Daily*, Statistics Canada, June 8, 2005.

9. *Globe and Mail*, June 23, 2004.

10. Nordicity Group Ltd.

11. *The Economist*, October 28, 2006.

12. *The Canadian Music Industry – 2005 Economic Profile*.

13. October 20, 2006.

14. *The Daily*, Statistics Canada, March 31, 2005.

15. *Globe and Mail*, December 7, 2006.

16. *Toronto Star*, October 19, 2005.

17. At this writing, the CRTC is still discussing the terms of the television broadcasting aspect of this takeover and has expressed reservations about who will really have effective control, CanWest Global Communications or the American Goldman Sachs Group.

18. Claire McCaughey, *Comparisons of Arts Funding in Selected Countries*, October 2005.

19. *Globe and Mail*, June 27, 2007.

20. *Globe and Mail*, June 16, 2007.

21. *Globe and Mail*, May 14, 2007.

CHAPTER TWENTY-THREE

1. March 3, 2005.

2. www.qnn.tv/threads, July 28, 2006.

3. Apparently this is now up to five.

4. *The Record* (Kitchener-Waterloo), and the *Guelph Mercury*, July 12, 2005.

5. Ben Bagdikian, *The Media Monopoly* (Boston: Boston Beacon Press, 2000).

6. *Literary Review of Canada*, January/February 2007.

7. *The Daily*, Statistics Canada, December 6, 2006.

8. *Globe and Mail*, April 5, 2007.

9. *USA Today*, September 24, 2007.

10. *New York Times,* September 16, 2006.

11. February 8, 2007.

12. *Globe and Mail,* March 21, 2007.

13. *Globe and Mail,* April 28, 2007.

14. January 25/February 1, 2007.

15. *Globe and Mail,* September 15, 2007.

16. *National Post,* March 14, 2007.

PART SIX

CHAPTER TWENTY-FOUR

1. *Investment Review,* Industry Canada quarterly.

2. *The Daily,* Statistics Canada, August 30, 2007.

3. The United Nations Conference on Trade and Development, October 2007.

4. *The Current,* CBC Radio, June 14, 2007.

5. *Financial Post,* June 14, 2007.

6. *Globe and Mail,* February 2007.

7. *Globe and Mail,* June 27, 2007.

8. *National Post,* May 9, 2007.

9. *Globe and Mail,* August 28, 2006.

10. Leonard Farber, Ogilvy Renault LLP, *Globe and Mail,* April 4, 2007.

11. *Financial Post,* June 4, 2007.

12. *Globe and Mail,* July 1, 2007.

13. Letter from Chamber of Commerce president and former Conservative cabinet minister Perrin Beatty to Industry Minister Jim Prentice, November 16, 2007.

14. *Globe and Mail,* April 17, 2007.

15. *Globe and Mail,* September 26, 2006.

16. *National Post,* March 31, 2007.

17. *Globe and Mail,* June 21, 2007.

18. *Globe and Mail,* May 2007.

19. *Globe and Mail,* April 13, 2007.

20. *Where I Stand* (Toronto: McClelland & Stewart, 1983).

21. *OECD Factbook,* 2006.

22. Latest development in the Canadian economic accounts, 13–605 XIE.

23. *Globe and Mail,* August 22, 2007.

24. Eric Beauchesne, *Financial Post,* June 22, 2007.

25. If you include the numbers from 2001 to 2006 inclusive, there were even more foreign takeovers – 1,825, according to Investment Canada.

26. Editorial, October 12, 2007.

27. For future reference, here's the number: (613) 954–1983.

28. Memo to his supporters, June, 2007.

29. *Canadian Dimension,* August 2, 2007.

30. CanWest News Service, July 14, 2007.

CHAPTER TWENTY-FIVE

1. Doug Davis, money manager at Davis-Rea Ltd., Toronto, *Toronto Star,* May 8, 2007.

2. One recent Ipsos Reid poll showed that 56 percent of Canadians believe that foreign ownership is a problem for Canada, while only 16 percent said it is not a problem.

3. April 2007.

4. *Toronto Star,* October 29, 2006.

5. August 21, 2006.

6. *Globe and Mail,* October 9, 2007.

7. *Globe and Mail,* March 30, 2007.

8. *Globe and Mail,* February 20, 2006.

9. *Globe and Mail,* May 4 and May 5, 2007.

10. March 11, 2007.

11. *Globe and Mail,* May 12, 2007.

12. *National Post,* April 9, 2007.

13. David Olive, *Toronto Star,* May 13, 2007.

14. May 8, 2007.

15. *Toronto Star,* October 29, 2006.

16. *Globe and Mail,* April 5, 2007.

CHAPTER TWENTY-SIX

1. *The Daily,* Statistics Canada, May 24, 2006.

2. *Canadian Economic Observer,* Historical Statistical Supplement, 2006/07, Statistics Canada.

CHAPTER TWENTY-SEVEN

1. January 25, 2007.

2. June 18, 2007.

3. *Globe and Mail,* November 6, 2006.

4. James Milway, *Maclean's,* March 19, 2007.

5. Statistics Canada table no. 3800017.

6. *Aggregate Productivity Measures,* Statistics Canada.

7. *OECD Factbook,* 2007.

8. *The Daily,* Statistics Canada, May 11, 2007.

9. Canadian Center for Policy Alternatives.

10. *BEA International Economic Accounts,* and Andrew Sharpe's CSLS tables.

CHAPTER TWENTY-EIGHT

1. August 28, 2006.

2. November, 2006.

3. February 2003.

4. *The Economist,* December 3, 2005.

5. *Toronto Star,* August 27, 2005.

6. *Globe and Mail,* September 17, 2005.

7. *Globe and Mail,* October 17, 2006.

8. *Globe and Mail,* August 20, 2005.

9. February 17, 2004.

10. *Globe and Mail,* September 29, 2005.

11. See Erin Weir, "Chapter Six, Canada's Free-Trade Agreements with the U.S. and Mexico: The Exaggeration of North American Trade," in *International Trade and Neoliberal Globalism,* Paul Bowles et al., eds (London: Routledge, 2008).

12. *Globe and Mail,* September 5, 2005. Spector is a former chief of staff to Prime Minister Brian Mulroney.

13. *Maclean's,* November 27, 2005.

14. *Globe and Mail,* December 23, 2006

15. *Toronto Star,* August 22, 2005

16. *Intent for a Nation* (Vancouver: Douglas & McIntyre, 2007).

17. *The Global Class War* (Toronto: John Wiley, 2006).

CHAPTER TWENTY-NINE

1. *The Economist* magazine's *Pocket World in Figures,* 2007.

2. *OECD Economic Outlook,* December 2006.

3. "Chained" refers to a technical method by which the National Accounts create inflation-adjusted data.

4. *Canadian Economic Observer,* Historical Statistical Supplement, 2006/07, Statistics Canada, p. 7.

5. *The Daily,* Statistics Canada, February 13, 2007.

6. *Canadian Economic Observer,* March 2006.

7. *OECD in Figures,* 2006, 2007.

8. *Globe and Mail,* November 10, 2006.

9. New Statistics Canada figures show a small surplus of $2.76-billion in automobile products in 2006.

10. *Toronto Star,* March 25, 2005.

11. *OECD Economic Outlook,* June 2007.

12. *Canadian Economic Observer,* Statistics Canada, May 2005.

13. *Globe and Mail,* March 2, 2006.

14. See "Cyclical Implications of the Rising Import Content in Exports," *Canadian Economic Observer,* Statistics Canada, 2002, and see Grant Cameron and Philip Cross, "The Importance of Exports to GDP and Jobs," *Canadian Economic Observer,* Statistics Canada, November 1999.

15. Published in 2006 by the Canadian-American Center at the University of Maine (www.umain.edu/canam).

16. *OECD Economic Outlook,* December 2006.

CHAPTER THIRTY

1. *Fair Trade For All* (New York: Oxford University Press, 2005).

2. *New York Times,* April 2, 2006.

3. *The Collapse of Globalism: And the Reinvention of the World* (Toronto: Viking Canada, 2005).

4. Review of Jeff Faux, *The Global Class War,* March 5, 2006.

5. *New York Times,* March 5, 2006.

6. *Princeton Progressive Review,* March 29, 2005.

7. September 2, 2006.

8. *OECD in Figures,* 2007.

9. *One Big Party* (Toronto: Hushion House, 2003).

PART SEVEN

CHAPTER THIRTY-ONE

1. Save the Children, 2007.

2. April 11, 2005.

3. *The Walrus,* October 2005.

4. *Toronto Star,* August 3, 2006.

5. *Global Child Survival and Health.*

6. *Maclean's,* November 20, 2006.

7. *The Social Benefits and Economic Costs of Taxation,* Canadian Centre for Policy Alternatives, 2007.

8. *The Economist,* May 28, 2005.

9. *Financial Times,* January 5, 2006.

10. *Globe and Mail,* April 22, 2005.

11. *Toronto Star,* May 8, 2007.

12. United Nations, *Human Development Report,* 2005.

13. United Nations, *Human Development Report,* 2005.

14. Reliable estimates say that inexpensive insecticide-treated bed nets can reduce malaria transmission by at least 50 percent.

15. United Nations, *Human Development Report,* 2005.

16. April 12, 2005.

17. Luiz Inácio da Silva.

18. Stephen Lewis, *Race Against Time* (Toronto: House of Anansi, 2005).

19. David R. Boyd, "Canada Is Failing the World's Poor," *Toronto Star*, May 8, 2007. David Boyd is an environmental lawyer, author, and Trudeau Scholar at the University of British Columbia.

20. *Globe and Mail,* June 5, 2007.

21. *Globe and Mail,* June 6, 2007.

22. Stephen Lewis, *Race Against Time* (Toronto: House of Anansi Press, 2005).

CHAPTER THIRTY-TWO

1. The American submarine fleet is all nuclear, but the Russians still have diesel/electric subs not unlike the subs Canada purchased from Britain.

2. In 2006, Canada became the 13th highest military spender in the world.

3. *NATO-Russia Compendium of Financial and Economic Data Relating to Defence* (1985–2006).

4. *Toronto Star,* February 13, 2007.

5. Angus Reid, June 11, 2007.

6. Strategic Counsel, *Globe and Mail,* July 13, 2007.

7. *Toronto Star,* October 27, 2007.

8. United Nations, *Human Development Report,* 2006, and the Stockholm International Peace Research Institute.

9. January 15, 2007.

10. *Vancouver Sun,* November 10, 2006

11. CBC TV, November 5, 2006.

12. June 29, 2006.

13. Steven Staples in *Marching Orders,* Council of Canadians.

14. *Maclean's,* November 20, 2006.

15. Decima, October 2, 2006.

16. February 24, 2007.

17. *Globe and Mail,* February 28, 2007.

18. *Toronto Star,* August 20, 2007.

19. James Ridgeway, *Counterpunch.*

20. *Globe and Mail,* July 17, 2007.

21. *New York Times,* February 26, 2007.

22. Steve Connor in the *Independent* (U.K.).

23. *Toronto Star,* October 27, 2006. Michael Byers holds the Canada Research Chair in Global Politics and International Law at the University of British Columbia.

24. Michael Byers, *Intent for a Nation* (Vancouver: Douglas & McIntyre, 2007), p. 158.

25. The 2004 Speech from the Throne promised to tackle the issue of the Northwest Passage once and for all.

26. September 10, 1985.

27. August 12, 2007.

28. June 14, 2007.

PART EIGHT

CHAPTER THIRTY-THREE

1. *OECD Economic Outlook,* June 2007.

2. *Fiscal Reference Tables,* Department of Finance, September 2007.

3. *Perspectives on Labour and Income,* Statistics Canada, Spring 2007.

4. "The Social Benefits and Economic Costs of Taxation," 2007.

5. From 1982 to 2005, transfers to persons were over 5 percent of GDP 10 times, over 4 percent 13 times.

6. *Globe and Mail,* June 16, 2006.

7. Earth Policy Institute, Eco-Economy Update, December 14, 2006–12, "Santa Claus Is Chinese" or "Why China Is Rising and the United States Is Declining."

8. April 18, 2005.

CHAPTER THIRTY-FOUR

1. This adds up to more than 100 percent because of transfers from Ottawa to the provinces and local governments, and transfers by the provinces to local governments.

2. *Toronto Star,* November 24, 2006.

3. *Globe and Mail,* May 2, 2007.

4. Ibid.

5. *Toronto Star,* March 31, 2007.

6. October 1, 2007.

7. August 14, 2007.

8. *Toronto Star,* September 27, 2007.

CHAPTER THIRTY-FIVE

1. *Globe and Mail,* April 25, 2007.

2. *Globe and Mail,* May 28, 2007.

3. Canadian Association of Petroleum Producers.

4. *The Daily,* Statistics Canada, February 28, 2007.

5. According to, for instance, *Petroleum Review* editor Chris Skrebowski. See David Crane, *Toronto Star,* July 15, 2005.

6. *Globe and Mail,* November 7, 2007.

7. *National Post,* March 6, 2007.

8. *Canada's Energy Outlook.*

9. *Globe and Mail,* October 5, 2006.

10. *Globe and Mail,* May 28, 2005.

11. Telephone conversation, January 10, 2006.

12. *Globe and Mail,* May 11, 2007.

13. One barrel of oil-sands oil creates three times as much GHG as one barrel of conventional oil.

14. *Globe and Mail,* May 23, 2007.

15. *Globe and Mail,* June 26, 2007.

16. *Toronto Star,* October 5, 2006.

17. *The Economist,* August 12, 2006.

18. "Selling the Family Silver," Parkland Institute, Edmonton, November 2006.

19. Strategic Counsel, *Globe and Mail,* September 16, 2006.

20. *Report on Business,* June 2006.

21. March 10, 2007.

22. Amy Taylor, Pembina Institute economist, Calgary, May 22, 2007.

23. February 17, 2005.

24. *National Post,* October 18, 2005.

25. *Globe and Mail,* March 9, 2007.

26. *Globe and Mail,* June 5, 2007.

27. *Globe and Mail,* September 19, 2007.

28. *Globe and Mail,* September 29, 2007.

29. *Financial Post,* October 2, 2007.

30. *Globe and Mail,* September 7, 2007.

31. *Globe and Mail,* October 28, 2007.

32. See William Marsden, *Stupid to the Last Drop* (Toronto: Knopf Canada, 2007).

CHAPTER THIRTY-SIX

1. *Globe and Mail,* October 2006.

2. For water usage see *OECD Factbook,* 2007, p. 165.

3. *OECD in Figures,* 2005, Supplement 1.

4. *Lies, Damned Lies and Trade Statistics,* Canadian-American Center at the University of Maine, 2006 (*www.umain.edu/canam*).

5. McClelland & Stewart, 2007.

6. "Water and NAFTA: The Politics of Change," *Canadian Organic Grower,* June 2004.

7. "International Economic Law: Water for Money's Sake?" (paper presented at conference in Brasilia, Brazil, September 2004).

8. Comments by Mel Clark to the author, April 2007.

9. *Globe and Mail,* November 11, 2005.

CHAPTER THIRTY-SEVEN

1. *OECD in Figures,* 2007.

2. *OECD Economic Outlook,* Statistical Annex, 2006.

3. *The Economist,* April 1, 2006.

CHAPTER THIRTY-EIGHT

1. *Globe and Mail,* June 29, 2007.

2. *New York Times,* July 29, 2005.

3. *Maclean's,* December 3, 2007.

4. The Innovative Research Group, *Maclean's,* March 19, 2007.

5. *Society at a Glance,* 2006, OECD.

6. Alberto Isgut, Lance Biales, and James Milway in *International Productivity Monitor,* Fall 2006.

7. *OECD Factbook,* 2007.

8. *OECD Factbook,* 2007.

9. *Insights on the Canadian Economy,* Statistics Canada, August, 2007.

10. *Globe and Mail,* November 9, 2006. The Center for Economic and Policy Research said in May 2007 that in France 30 days of legally required paid leave was the standard. This compares with 25 days in Sweden, 22 in Spain, 20 in Australia, Germany, and the United Kingdom, 10 in Canada and Japan, and zero in the United States.

11. *Globe and Mail,* November 19, 2007.

12. *Globe and Mail,* November 23, 2007, based on *The Daily,* Statistics Canada, November 22, 2007.

13. *Toronto Star,* November 25, 2007.

14. Letter to the editor, *Globe and Mail,* November 20, 2007.

PART NINE

CHAPTER THIRTY-NINE

1. *Society at a Glance,* 2006, OECD.

2. *Globe and Mail,* October 20, 2007.

3. To be exact, 174,661 people made donations out of a total of 23,054,614 eligible voters, for a percentage of 0.76.

4. Fair Vote Canada, January 24, 2006.

5. 1940, 1949, 1958, 1984.

6. The 2006 figures were based on $1.75 per vote.

7. In the first half of 2007, 45,192 Canadians contributed to the Conservative Party, 14,782 to the NDP, only 4,367 to the Liberals, 2,669 to the Green Party, and 476 to the Bloc Québécois.

8. *Toronto Star,* January 3, 2007.

9. *Globe and Mail,* March 22, 2007.

10. *Policy Options,* September 2005.

11. A fuller description of the workings of an MMP system can be found on Fair Vote Canada's website: www.fairvotecanada.org. The Law Commission of Canada's report is titled *Voting Counts: Electoral Reform in Canada* (www.lcc.gc.ca, or info@lcc.gc.ca, or 613–946–8980).

12. An excellent source of comparative international data appears in Arend Lijphart, *Patterns of Democracy, Government Forms and Performance in Thirty-Six Countries* (New Haven: Yale University Press, 1999). Fair Vote Canada also has a 10-page summary of key findings: "Can Fair Voting Systems Really Make a Difference?" available on the Publications page at www.fairvote.ca.

13. *CCPA Monitor,* September 2004.

14. For more information on the single transferable vote (STV), see www.citizensassembly.bc.ca.

15. *Georgia Straight,* February 10, 2005.

16. April 14, 2007.

17. *Globe and Mail,* October 1, 2007.

CHAPTER FORTY

1. United Nations, *Human Development Report,* 2005, and *The Economist* magazine's *Pocket World in Figures,* 2007.

2. *Toronto Star,* November 26, 2007.

3. *The Daily,* Statistics Canada, November 7, 2006.

4. *Education at a Glance,* 2006, OECD.

5. *The Daily,* Statistics Canada, December 8, 2006.

6. *The Daily,* Statistics Canada, November 23, 2006.

7. *The Daily,* Statistics Canada, January 5, 2007, and *Canadian Economic Observer,* Statistics Canada, June 2007.

8. *Society at a Glance,* 2007, OECD.

9. *The Daily,* Statistics Canada, March 7, 2006.

10. *Globe and Mail,* January 29, 2007.

11. *OECD Factbook,* 2007.

12. *The Daily,* Statistics Canada, September 24, 2004.

13. *Literary Review of Canada,* April 2007.

14. United Nations, *Human Development Report,* 2006.

15. United Nations, *Human Development Report,* 2005.

CONCLUSION

1. October 3, 2002.

2. *Globe and Mail,* September 15, 2001.

3. Ibid.

4. Bravo to Aaron Paton of the *Banff Crag & Canyon,* who won a Community Newspaper Association award for his efforts to properly publicize the story.

5. Elliott was appointed as the new RCMP commissioner by Stockwell Day in July 2007.

6. *Ottawa Citizen,* March 23, 2007.

7. E-mail, March 31, 2007.

8. In September 2007, an American SPP negotiator, Joseph Dukert, an adjunct fellow at the U.S. Center for Strategic and International Studies, warned that the continentalism of energy is already so well entrenched that "it would be a provocation if a future government . . . tried to back out of the SPP."

9. Democrat Marcy Kaptur.

10. *Globe and Mail,* July 9, 2007.

11. *National Post,* August 29, 2007.

12. Testimony before the House of Commons Standing Committee on International Trade (CIIT) Canada-U.S. Trade and Investment Issues and the SPP.

13. Ipsos Reid poll, conducted for CanWest News and Global News, October 16, 2007.

GLOSSARY

1. *OECD Factbook,* 2006.

APPENDIX ONE

1. Judy Foreman, February 10, 2004.

2. *Globe and Mail,* July 13, 2005.

3. *New York Times,* December 31, 2006.

4. *Toronto Star,* March 21, 2006.

5. *New York Times,* June 16, 2004.

6. Economic Policy Institute, Washington, D.C., 2006.

7. *New York Times Book Review,* March 23, 2006.

8. *New York Times,* March 2, 2007.

9. *Time,* May 29, 2006.

APPENDIX TWO

1. *The Daily,* Statistics Canada, September 6, 2005.
2. *Toronto Star,* April 15, 2007.
3. See, for example, the *Post's* January 15, 2007, lead editorial.
4. *Child Poverty in Rich Countries,* 2005.
5. March 12, 2006.
6. *The Province,* February 25, 2007.